# Fuzzy-Neural Control

# Fuzzy-Neural Control

## Principles, algorithms and applications

**Junhong Nie and Derek A. Linkens**

**Prentice Hall**

New York London Toronto Sydney Tokyo Singapore

First published 1995 by
Prentice Hall International (UK) Limited
Campus 400, Maylands Avenue
Hemel Hempstead
Hertfordshire, HP2 7EZ
A division of
Simon & Schuster International Group



Printed and bound in Great Britain by
Bookcraft Ltd, Midsomer Norton, Avon

Library of Congress Cataloging-in-Publication Data

Nie, Junhong.
Fuzzy-neural control : principles, algorithms, and applications / by Junhong Nie and Derek A. Linkens.
p. cm.
Includes bibliographical references and index.
ISBN 0-13-337916-7
1. Neural networks (Computer science) 2. Fuzzy systems.
I. Linkens, D.A. II. Title
QA76.87.N498 1995
629.8'9--dc20 94-47246
CIP

British Library Cataloguing in Publication Data

A catalogue record for this book is available from the British Library

ISBN 0-13-337916-7

1 2 3 4 5 99 98 97 96 95

# Contents

# Preface

Reflecting the authors' recent research results in the area of fuzzy and neural systems and learning control, this book provides a unified and comprehensive treatment of fuzzy and neural control systems under the frameworks of knowledge representation, reasoning, and acquisition which are recast techniquely into more practical issues, being appropriately termed as structure, computation, and learning for use in intelligent control. Considerable effort is devoted to developing systems which are capable of performing self-organizing and self-learning functions in a real-time manner under multivariable system environments by utilizing fuzzy logic, neural networks, and a combination of both paradigms with emphasis on the novel system structures, algorithms, and applications to some problems found in industrial and biomedical systems.

The book essentially consists of two independent but closely correlated parts. The first part (Chapters 2-5) involves mainly the utilization of fuzzy algorithm-based schemes including topics such as approximate reasoning models, multivariable fuzzy controller structures, and learning algorithms, whereas the second part (Chapters 6-10) deals primarily with the subjects of fuzzy-neural network-based approaches containing four hybrid fuzzy-neural controller structures with corresponding self-learning and self-organizing algorithms, covering five fundamental and widely used network paradigms including BNN, CPN, CMAC, RBF, and LVQ networks. Thus, the book:

1. Provides a unified and comprehensive treatment of fuzzy algorithm-based and neural network-based control systems under the framework of knowledge representation, reasoning, and acquisition.
2. Offers systematic methodologies to merge fuzzy systems with neural networks for use in intelligent modelling and control.
3. Presents a unified approximate reasoning model suitable for embedding linguistic definitions, for handling uncertainties, and for dealing with linguistic and numerical environments.
4. Introduces four novel fuzzy-neural controller structures with related reasoning algorithms and detailed implementations.

5. Provides four new self-learning and self-organizing strategies to construct controllers systematically with solid mathematical foundations for the learning mechanism.
6. Demonstrates the feasibility of the proposed approaches by applying them to the problem of multivariable blood pressure control.

## Acknowledgements

We would like to express our sincere thanks to Professor E. H. Mamdani for his valuable comments on a draft of the manuscript, and to Professor C. J. Harris for providing relevant references which helped to improve the completeness of the book. We are much indebted to Dr M. Mahfouf who has done an excellent job in editing the book.

Junhong Nie would also like to thank Professor C. C. Hang and Dr T. H. Lee at National University of Singapore for their support which made it possible for him to have time to complete the book. Thanks also go to the programmers A. J. Cornah and L. S. Gray, who have provided constant assistance in the use of the Sun Workstations when he was in the Department of Automatic Control and Systems Engineering, the University of Sheffield, UK. Last, but not least, he would like to thank his wife, Dongju, and his daughter, Miao Nie for their patience, understanding and complete support throughout this work.

Sheffield
30 September 1994

Junhong Nie and Derek A. Linkens

CHAPTER 1

# Introduction

## 1.1 Expert systems, fuzzy sets and control theory

Despite being very controversial, rule-based fuzzy control has been and continues to be a very active and fruitful research field since Mamdani's pioneering work (Mamdani 1974) which was based on Zadeh's novel approach formulated in his seminal paper (Zadeh 1973). The impetus behind this continuity of the subject lies largely in the fact that numerous applications of fuzzy control have emerged covering a wide range of practical areas, and that many software and hardware products for fuzzy control (Schwartz and Klir 1992) have been commercialized during the last few years. A large body of literature on fuzzy control exists. Some comprehensive survey papers (e.g. Mamdani 1977, 1993; Tong 1984, 1985; Sugeno 1985; Maiers and Sherif 1985; Efstathiou 1989; Lee 1990) are helpful for a quick access to this field. Notice that here we have no intention to give a complete survey on fuzzy control, rather we are more concerned with presenting a general picture of the field and clarifying some fundamental issues related to it.

As an outcome of merging the techniques of traditional rule-based expert systems, fuzzy set theory, and control theory, fuzzy control departs significantly from traditional control theory which is essentially based on mathematical models of the controlled process. Instead of deriving a controller via modelling the controlled process quantitatively and mathematically, the fuzzy control methodology tries to establish the controller directly from domain experts or operators who are controlling the process manually and successfully. Clearly, this is a typical characteristic of an expert system where primary attention is paid to the human's behaviour and experience, rather than to the process being controlled. It is this distinctive feature that makes fuzzy control applicable and attractive for dealing with those problems where the process is so complex and ill-defined that it is either impossible or too expensive to derive a mathematical model which is accurate and simple enough to be used by traditional control methods, but the process may be controlled satisfactorily by human operators. However, we must be cautious when making the claim that fuzzy control does not need a process model, a claim which leads easily to a misunderstanding about fuzzy control. It would be impossible to control a process by either a human or a machine without knowing something about the process. In fact, all the knowledge about the process has been embedded implicitly into the control policy by the domain experts or operators.

Having recognized that human control behaviour is the basis for implementing fuzzy control, it is necessary to express the human knowledge in an easy and effective way. The IF-THEN rule format is one of the most suitable representations for control applications. The control knowledge can be expressed by a set of linguistic rules with the form of IF *situation* THEN *action*. It is noted that this kind of statement possesses two distinct features: it is qualitative rather than quantitative; it is a local knowledge associating a local situation with an appropriate action. The former can be characterized by fuzzy subsets, whereas the latter can be expressed by a fuzzy implication or a fuzzy relation. However, the ultimate form of operation must be numerical in accordance with the nature of current numerical computers. Fortunately, membership functions and approximate reasoning provide powerful means to numerically characterize fuzzy sets and fuzzy implication. It is the rule-based structure combined with fuzzy set theory that makes the implication of the fuzzy control possible. In fact, Zadeh's fuzzy set theory, possibility theory and approximate theory not only offer a theoretical basis for fuzzy control, but also establish a bridge connecting Artificial Intelligence to control engineering.

## 1.2 Representation, reasoning and acquisition

There are three fundamental issues concerning the implementation of a practical fuzzy control system. There are *knowledge representation*, *reasoning strategy*, and *knowledge acquisition*. Recognizing that a rule-based structure is adopted throughout this book, then the *representation* issue will have a specific meaning of how linguistic rules can be represented numerically. *Reasoning strategy* concerns the problem of how a reasonable conclusion (action) should be made with respect to an incoming situation. *Knowledge acquisition* deals with the issue of how a set of control rules can be derived. It should be pointed out that these three issues are closely related.

By noting that the knowledge provided by experts or operators in a specific domain is usually qualitative and contains some kind of uncertainty, it is inevitable that both knowledge representation and reasoning must be fuzzy or approximate. This has resulted in the emergence of a theory known as Approximate Reasoning (AR) (Gupta *et al.* 1985) which can be viewed as a process by which a possible imprecise conclusion is deduced from a collection of imprecise premises (Zadeh 1973). Some researchers, for instance Zadeh (1983), Yager (1984) and Turksen and Zhong (1988), have discussed the issues of fuzzy representation and approximate reasoning in general expert systems. Others, for example Mamdani (1974), Mamdani and Assilian (1975), Holmblad and Ostergaard (1982) and Sugeno and Tagaki (1983), have developed some approximate reasoning algorithms from a real-time control point of view. More recently, it has been demonstrated that approximate reasoning schemes can be realized in chips using commercially available VLSI hardware devices (Lim and Takefuji 1990; Togai and Watanable 1986). Among these algorithms, the one widely used in real-time experts systems is the *relational matrix* model, which was first adopted by Mamdani and is based on Zadeh's Compositional Rule of Inference (Zadeh 1973). Because rules can be interpreted as implication relations in a logical sense, numerous implication operators have been suggested in the literature (Mizumoto and Zimmermann 1982). In addition, following Zadeh's inference idea, several other composition approaches have been proposed in order to achieve certain desired properties (Pedrycz 1985a).

Knowledge acquisition, being thought of as a bottleneck for building a realistic expert system, plays a crucial role in designing fuzzy control systems. Basically, there are two approaches for building the rule-base, namely, either from human experts and/or from the process being controlled. The former method, which is common in general expert systems, has a heuristic and qualitative nature. Expert experience and control engineering knowledge are extracted, either through interrogation of experienced experts or through on-line identification of skilled operators' control actions (Sugeno 1985), as a set of IF-THEN rules relating the measured variables to control variables. It is evident that this approach relies on the availability of domain experts. When there are no experts or skilled operators available at hand to supply necessary knowledge or to be used as an identified model, it would be desirable to construct the rule-base by directly operating the process being controlled. This constitutes a very challenging and important topic which will be treated intensively in the discussion which follows.

## 1.3 Inference engines and function approximator

As mentioned above, to implement a fuzzy control system, three issues in terms of knowledge-based expert systems are normally involved, which are knowledge representation, reasoning, and acquisition. This section will address the first two issues further from two different angles and the last one, knowledge acquisition, will be the topic of Section 1.4.

### 1.3.1 Fuzzy systems as inference engines

Assume that, for a specific system, a set of inputs $X$ and outputs $Y$ are appropriately identified and, for a specific application, a rule-base $\Psi_X^Y$ is derived from available sources containing a set of rules relating $X$ to $Y$ by using predefined linguistic labels. With the aid of fuzzy set and fuzzy logic theory, a fuzzy system at a linguistic level can be established as shown in Figure 1.1. For a current input $X_0$, the corresponding output $Y_0$ is inferred as $Y_0 = \Phi(\Psi_X^Y, X_0)$. The mechanism of the system in deriving a reasonable action $Y_0$ with respect to a specific situation $X_0$ can be interpreted simply as performing a two-stage reasoning process. Viewing the rule-base $\Psi_X^Y$ as a prototype, the inference engine $\Phi$ first carries out a *matching* procedure between the current situation $X_0$ and the IF parts of the rules. To deduce the corresponding action, then an *interpolative* procedure takes place by operating the matching results with the THEN parts of the rules using a fuzzy inference strategy.

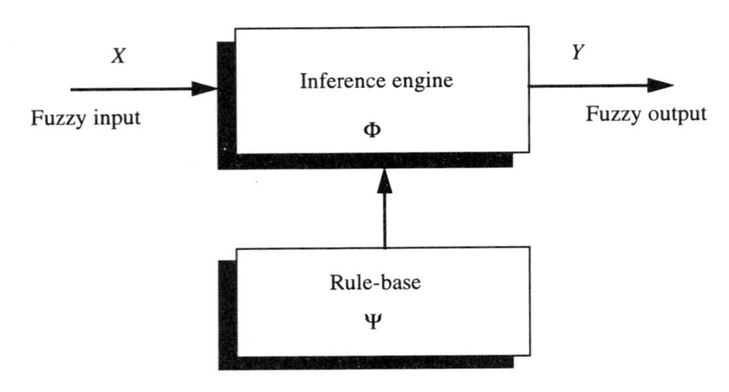


**Figure 1.1** Fuzzy system as inference engine

There are two important points worth noticing. First, apart from the dependence on matching and inference methods being used, the performance of the system rests largely on the quality of the rule-base. An unreliable and poorly formed rule-base is clearly unlikely to perform the task satisfactorily. Second, the system inherently possesses an interpolative or approximate reasoning function. More specifically, for an already known situation $X^*$, the inferred action $Y$ should correspond to an already known $Y^*$ exactly or at least satisfy, in some meaningful sense, $|Y - Y^*| < \varepsilon$ where $\varepsilon$ is a small number and $(X^*, Y^*)$ is an IF-THEN pair which exists in the rule-base. On the other hand, for a novel situation $X_0$ to which there is no rule corresponding exactly, an approximate action $Y_0$ can still be deduced by the inference engine. Thus, a fuzzy system may be viewed as a linguistic interpolative table following a simple rule: producing a similar action in response to a similar situation.

To implement a fuzzy system in a computational form, one of the methods used widely in the fuzzy community is to construct a relation matrix $R$ from the available rule-base and then the current output $Y$ is calculated by a relation equation $Y = X \bigcirc R$, where the symbol $\bigcirc$ denotes a logical operator performing composition of inference. From a system's viewpoint, the relation equation provides a compact formulation, analogous, for example, to the convolution equation in linear systems theory, with the relation matrix as the counterpart of the impulse response function and the composition operator corresponding to the convolution operator. It had been hoped that by this elegant representation or relation equation a solid fuzzy system theory could be established in a mathematically vigorous way by introducing many fundamental concepts found in well-established linear systems theory into fuzzy systems. Unfortunately, except for some conceptual and formal descriptions concerning fuzzy systems, there have not been useful achievements and conclusions which provide practical assistance for fuzzy system applications, particularly in engineering fields.

### 1.3.2 Fuzzy systems as function approximators

The above discussion assumes that the working environment of the fuzzy system is of a linguistic nature, meaning that situations $X$ and actions $Y$ assume linguistic labels as their values which are represented by corresponding fuzzy subsets with suitably chosen membership functions. However, the majority of engineering problems involve only numerical variables, such as in the case of control and modelling. To apply the fuzzy system to those numerical environments, two procedures, namely fuzzification and defuzzification, are normally employed as shown in Figure 1.2. The measured values $x$ are converted into fuzzy values $X$ compatible with the form of the fuzzy system. The inferred values $Y$ have to be converted into crisp values $y$ compatible with the form of the numerical environment. Thus, we realize that although the fuzzy system itself has an association at a set level, $Y{=}F(X)$, mapping a fuzzy set $X$ into a fuzzy set $Y$, the ultimate mapping from $x$ to $y$ in fact represents a determined function $y = f(x)$ mapping a specific $x \in \bar{X}$ into $y \in \bar{Y}$. In this sense, the augmented fuzzy system of Figure 1.2 can be considered as a numerical function realization which is determined mainly by the rule-base and associated inference algorithm. More precisely, the fuzzy system is a function approximator due to its interpolative nature as discussed previously. Thus, viewing the fuzzy system as a black-box with corresponding representation and reasoning schemes, we can use the constructed fuzzy system to represent a known or unknown function. Notice that the function to be represented can be nonlinear. This point will be developed further in the discussion which follows.

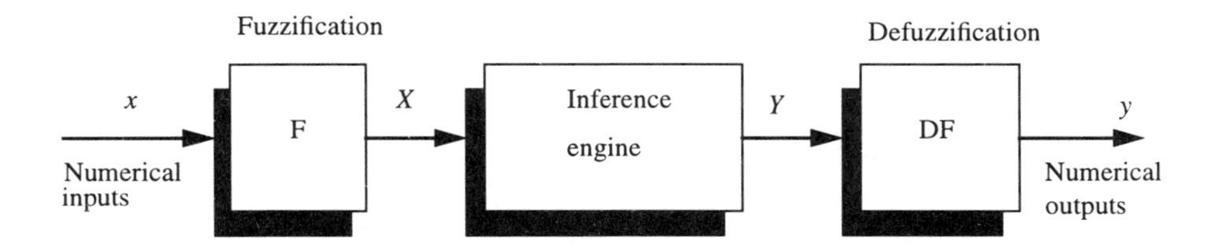


**Figure 1.2** Fuzzy system as function approximator

### 1.3.3 Numerical level versus linguistic level

From practical and implementational standpoints, it is worth while to distinguish two different environments within which a fuzzy system may be involved: developing an environment in which the fuzzy system is designed off-line, and a working environment in which the designed fuzzy system is actually operated. Traditionally, a fuzzy system is built by using linguistic rules derived normally from human experts and fuzzy set theory, and thus it can readily work in a linguistic environment as an inference engine at a set level shown in Figure 1.1. However, when applying such a developed system to a numerical environment, two interfaces as indicated in Figure 1.2 have to be added. Here, we implicitly employ a philosophy of *making the working environment which is numerical, adapt to the developing environment which is linguistic.* Membership functions play a central role in converting qualitative linguistic labels into quantitative graded values. Considerable computational difficulties arise in adopting the above philosophy, particularly when dealing with multivariable systems. To see this, assume that the fuzzy system has $m$ inputs $X_1, X_2, \ldots, X_m$ and a single output $Y$, $X_i$ and $Y$ being defined over universes of discourse $\overline{X}_i$ and $\overline{Y}$ respectively. To construct a relation matrix $R$, each universe has to be discretized as $\overline{X}_i = x_{i1}, x_{i2}, \ldots, x_{ini}$ and $\overline{Y} = y_1, y_2, \ldots, y_{ny}$, over which corresponding fuzzy sets are defined. In this regard, the fuzzy system performs a function mapping at a set level from an $l$-dimensional unit cube space $I^l$ to a $k$-dimensional unit cube $I^k$ (Kosko 1992b), where $l = n_1 + n_2 + \ldots + n_m$ and $k = n_y$. In practice, $n_i$ and $y_n$ are normally more than 13 and thus the relation matrix $R$ will be a $(m + 1)$-dimensional matrix with $n_1 n_2 \ldots n_m$ elements. For a moderate $m = 4$ and $n_i = y_n = 13$, we have $l = 52$, $k = 13$, and a 5-dimensional matrix $R$ with $13^4 = 28\,561$ elements. It is evident that such a high-dimensional operation can very easily discourage any attempt to apply such a method to a system with multivariable inputs, although $R$ may be precalculated as a look-up table when the number of variables involved are low enough, for example 2.

The question naturally arises as to whether it is possible to utilize an alternative philosophy in implementing a fuzzy system, that is, *to make the developing environment adapt to the working environment* instead of vice-versa as used above. If this is the case, the fuzzification and defuzzification procedures should be incorporated directly into the reasoning mechanism with possibly different conceptual interpretation, and they would not be explicitly needed in the working mode since the developed fuzzy system is capable of accommodating the numerical environment automatically. Thus, the working fuzzy system involves only numerical quantities in the input product space $\overline{X}_1 \times \overline{X}_2 \times \cdots \times \overline{X}_m$ with a low dimension $m$ instead of $l$ and one-dimensionial output space instead of $k$. The above arguments and considerations may not only bring about the possibility of developing more efficient fuzzy systems but also may lead to easier integration of fuzzy systems with other approaches, particularly learning systems and neural networks which work at a numerical level as will be detailed in several chapters of this book.

## 1.4 Model-based and learning-based fuzzy control

As mentioned in Section 1.1 fuzzy control can be regarded as a system integrating the concepts of expert systems, control theory, and fuzzy theory. The development of fuzzy control can be naturally along two distinct lines: AI-based and control theory-based with fuzzy theory as a means of soft computing. In the former case, human control knowledge and experience play central roles in designing the control system, whereas in the latter more emphasis is put on human behaviour such as adaptive and learning capabilities which have something to do with control theory. This book will be concentrating on the latter case.

### 1.4.1 The role of fuzzy systems in control

Regarding a fuzzy system as either an approximate reasoner or a functional approximator, it can be utilized in various ways in control systems. The most popular and the simplest usage is to use it as a direct controller as shown in Figure 1.3. The second utilization of a fuzzy system is to use it as a supervisor as shown in Figure 1.4. Instead of directly issuing the control signals, the function of the fuzzy system lies in monitoring a low-level direct controller by outputting appropriate parameters to be used by the direct controller. The decision taken by the supervisor can be based on the current control performance or the operating conditions depending on the control strategies employed in the system. By treating the fuzzy system as a representative model of the controller plant, we can find the third usage of the fuzzy system, that is, to place it into the various model-based control structures found in traditional control systems by replacing the traditional model with a fuzzy one.

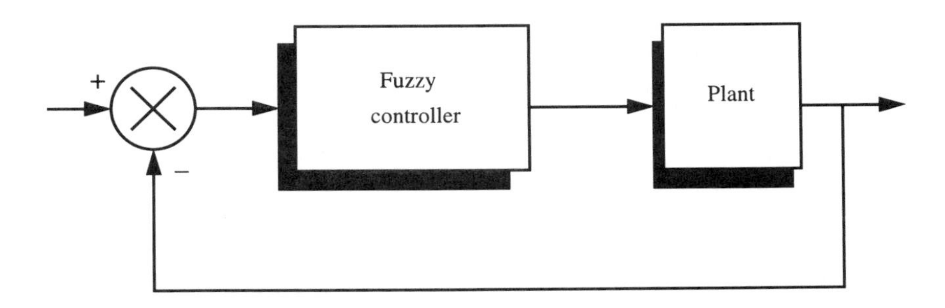


**Figure 1.3** A direct fuzzy control system

The following discussion will focus on the first stage of the fuzzy systems although we use the notions and even structures similar to those of the other two

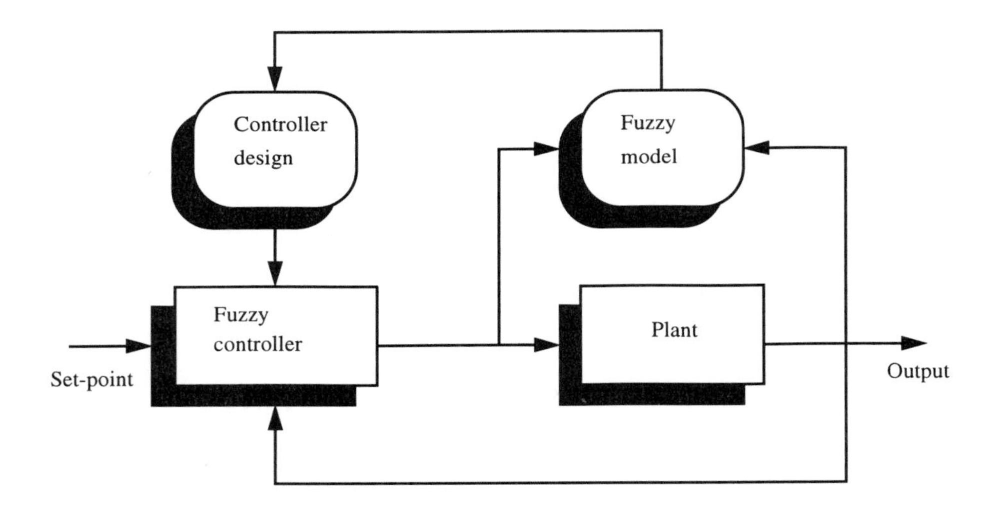


**Figure 1.4** Model-based fuzzy control

usages but with different interpretations. Referring to Figure 1.3, a basic requirement for implementing such a fuzzy controller is the availability of a control expert who provides the necessary knowledge for the present control problem. However, there are cases where either no expert exists or it is very difficult and expensive to derive control rules due, for example, to the complexity and the uncertainty of the controlled process. Under these circumstances, a suitable rule-base must be constructed automatically by directly operating the process being controlled, provided that one still wants to utilize the rule-based paradigm as the controller structure. The development of the research effort covered in this book has been mainly along two lines: model-based and learning-based. The former is more closely related to a traditional adaptive control strategy, whereas the latter has more connection with machine learning and intelligence. As will be seen, although both strategies can be considered as performing the function of knowledge acquisition, the former carries out this process implicitly, whereas the latter does it explicitly.

### 1.4.2 Model-based fuzzy control

The basic principle of this approach is that the relationship governing the process inputs and the outputs are assumed to be represented by a fuzzy system in terms of either a set of IF-THEN rules or a relational equation called a fuzzy model. First, an identification procedure is applied. After the fuzzy model is appropriately identified, a set of control rules or a controller relational equation can be created based on the identified model. Clearly, this is a well-known scheme based on the principle of *certainty equivalence* and adopted by traditional indirect adaptive systems where two stages, model identification and controller redesign, are involved. By viewing a fuzzy rule-based paradigm as defining a relation matrix, Pedrycz (1985b) proposed a methodology to derive such a matrix via solution of pre-identified relation equations governing the controlled process. Takagi and Sugeno (1985) developed an identification approach capable of extracting control rules from skilled operators, and they further claimed that the method could be used to identify process models in the form of rules and then to obtain control rules based on the identified model. Rhee *et al.* (1990) presented an alternative fuzzy control structure. The system contains a cell-like knowledge structure that represents the input–output relationship of the controlled process both in time and in value. The knowledge structure is established during the learning phase and is subsequently used to control the process based on two sets of inference rules. Moore and Harris (1992) described an indirect adaptive fuzzy controller as shown in Figure 1.4. A conventional fuzzy controller contains the same components as shown in Figure 1.1. However, the fuzzy relation $R$ of the controller is derived by an on-line design procedure using the concept of causality inversion of the fuzzy relation characterizing the process being controlled. The desired closed-loop performance specified by another fuzzy relation can also be included when designing the controller. Although the method is systematic and possesses many useful features, in practice it is limited to a single variable system. As discussed previously, due to involving high dimensional space, extending the method to multivariable cases not only creates tremendous computational effort in solving the relation equations but also introduces the need for dealing with interaction effects between the variables.

### 1.4.3 Learning-based fuzzy control

In contrast with the model-based approach, the learning-based methodology tries to emulate the human's learning ability by means of operating the process repeatedly, and thus the process behaviour is not explicitly taken into account. Here, past experience is of particular importance and is utilized intensively for creating workable control systems. A notable example belonging to this class is a so-called linguistic self-organizing controller (SOC) which was developed by Procyk and Mamdani (1979). The architecture of the SOC is illustrated in Figure 1.5. A performance feedback loop is added to a basic fuzzy controller. On-line measured performance of the control system indicated by a performance index table (PIT) is used to create or modify the rule-base. Starting from an empty rule-base, the SOC is able to construct a suitable rule-base via which the resulting control performance is acceptable in the sense of following the PIT measure. Thus, with little *a priori* knowledge about the process, the rule-base is self-organized with the learning process. It should be noted that in the original work of Procyk and Mamdani, the operation of the basic fuzzy controller was based on the relation equation method as shown in Figure 1.3. Therefore, the modified object was the controller relation matrix while created and/or modified rules were used as the means for forming such a matrix.

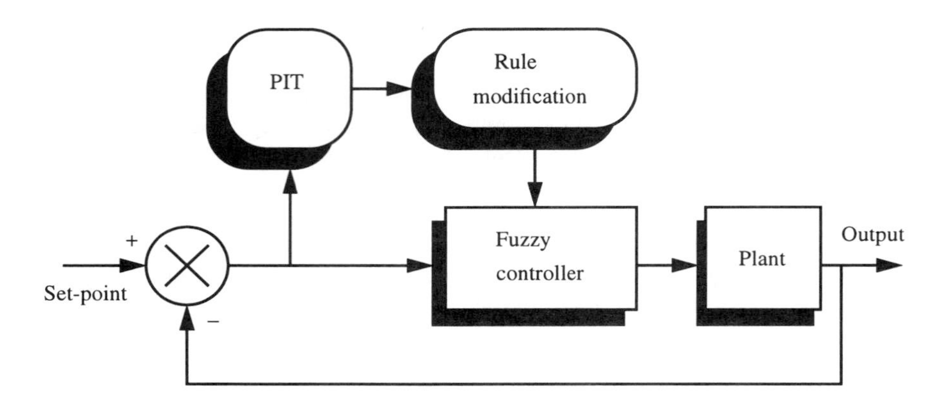


**Figure 1.5** Learning-based fuzzy control

Following the work of Procyk and Mamdani, the performance of SOC has been investigated intensively and some improved algorithms and design procedures as well as successful applications have been reported (e.g. Mamdani *et al.* 1984; a series of PhD theses at QMC (Yamazaki 1982; Lembessis 1984; Sugiyama 1986; and Tansheit 1988), Scharf and Mandic 1985; Daly and Gill 1986; Shao 1988; Linkens and Abbod 1992). The application fields include robotic control, process control, spacecraft attitude control, etc. The SOC has the important advantage of self-acquiring the required knowledge without relying on human experts. This in turn represents a substantial step forward for fuzzy control research. Like the model-based case, however, the SOC works in high-dimensional space even when the process itself is a single variable system and therefore, as would be expected, it has difficulty in handling multivariable systems.

## 1.5 Neural networks and fuzzy systems

It is generally recognized that the most salient features of a neural network, being biologically inspired or not, lie in its parallel structure consisting of massively connected computation units, its trainable property and its nonlinear mapping capacity. Numerous networks have been suggested and investigated with various topological structures, functionality, and training algorithms (Hecht-Nielsen 1990). For control applications, we are particularly interested in those which can perform functional approximations. Some well-known networks in this category include Back-propagation Neural Networks (BNN) (Rumelhart *et al.* 1986), Radial Basis Function (RBF) (Broomhead and Lowe 1988), Cerebellar Model Articulation Controller (CMAC) (Albus 1975a, b), and Counterpropagation Network (CPN) (Hecht-Nielsen 1987; Simpson 1990).

### 1.5.1 Knowledge processing: an alternative viewpoint

Although the origins and motivations of fuzzy systems and neural networks are quite different, with the former trying to capture human thinking and reasoning capability at a cognitive level and the latter attempting to mimic the mechanism of the brain at a biological level, they may be described by a common language, particularly when they are employed in solving engineering problems. From a knowledge processing viewpoint, many aspects of neural networks can be discussed similarly to fuzzy systems in terms of representation, reasoning and acquisition.

*Representation versus structure.* A trained neural network can be viewed as a means of knowledge representation. Instead of representing knowledge using IF-THEN localized associations as in fuzzy systems, a neural network stores knowledge through its structure, and more specifically its connection weights and local processing units in a distributed or localized manner.

*Reasoning versus computing.* Feedforward computing in neural networks plays the same role of forward reasoning as that in fuzzy systems. Both systems carry out the task by operating on the stored knowledge with respect to the current situation (input) so as to obtain a required action (output). Responding to a novel situation by giving an approximate action is a central point for both systems. However, the methods for completing the task are different. Fuzzy systems are logical-inference-based with interpolative reasoning, whereas neural networks are algebraic computing-based with a natural generalization capability.

*Acquisition versus learning.* Fuzzy systems acquire knowledge normally from domain experts and this knowledge is then programmed into the system with the aid of fuzzy logic theory. In contrast, neural networks acquire knowledge usually from samples, and the knowledge is encoded into the network by training. It should be emphasized that although the words training and learning are exchangeable in most of the neural networks literature, here the word learning is applied not only in the sense of training (a process of constructing a network with known samples) but also, more importantly, in a broader sense of constructing networks with samples which are not readily available. This point will be addressed further in the discussions that follow.

### 1.5.2 Universal approximation: a unified interpretation

Functionally, a fuzzy system or a neural network can be best described mathematically as a function approximator. More specifically, they aim at performing an approximate implementation of an unknown mapping $\phi : A \subset R^n \rightarrow R^m$, where $A$ is a compact of $R^n$, by means of available knowledge relevant to the unknown mapping. Theoretical investigations (e.g. Funahashi 1989; Hornic 1991; Hornic *et al.* 1989; Park and Sandberg 1991, 1993) have revealed that neural networks are in fact universal approximators. More precisely, these results show that standard network architectures such as BNN and RBF are capable of approximating virtually any function of interest to any degree of accuracy on a compact set $A$, provided that sufficiently many hidden units are available.

It is interesting to note that recently, motivated by the results of neural network researchers, several authors have shown a similar approximating property with respect to fuzzy systems. By showing that fuzzy systems are dense in the space of continuous functions, Kosko (1992a) and Wang (1992) (using different fuzzy system structures) independently proved that fuzzy systems can approximate arbitrarily well any continuous function on a compact domain. Buckley *et al.* (1993) argued that fuzzy systems and feedforward neural networks are able to approximate each other to any degree of accuracy, implying the universal approximate property of fuzzy systems. By using the equivalence principle between fuzzy and neural systems, Chapters 6 and 10 make a similar claim regarding the approximating ability of fuzzy systems.

### 1.5.3 Integration of fuzzy systems with neural networks

Traditionally, fuzzy systems and neural networks have been investigated along rather different lines due partly to the fact that they are derived from rather different fields. While a fuzzy system possesses great power in representing linguistic and structured knowledge by fuzzy sets and performing fuzzy reasoning by fuzzy logic in a qualitative manner, it usually relies on domain experts to provide the necessary knowledge for a specific problem. On the other hand, while neural networks are particularly good at representing nonlinear mappings in computational fashion, they construct themselves generally via training procedures presented with samples. Furthermore, while the behaviour of fuzzy systems can be understood easily due to their logical structure and step-by-step inference procedures, a neural network acts normally as a black-box, without providing explicit explanation facilities.

It is quite natural to consider the possibilities of integrating the two paradigms into a new kind of system where the desired strengths of both systems are utilized and combined appropriately. In fact, recently there has been a great amount of interest and practice in the synergetic combination of fuzzy systems and neural networks, with an expectation that the capacity of the hybrid system will be greatly enhanced. The state of the art in this area is largely reflected in recent monographs (Bezdek and Pal 1992; Harris *et al.* 1992; Kosko 1992b; Wang 1994), journal special issues (Harris *et al.* 1992; Bezdek 1992), and articles (e.g. Berenji 1992; Brown and Harris 1992; Carpenter *et al.* 1992; Lee 1991; Takagi and Hayashi 1991; Yamaguchi *et al.* 1992). We have not attempted to review all the work in depth, nor do we mean that the cited references are an exhaustive list. Interested readers are advised to consult the above references and references therein. However, many of the existing ideas and approaches to the integration of both systems, particularly those pertaining to control applications, will be addressed in some detail in the next section.

## 1.6 Fuzzy-neural control: ideas and paradigms

### 1.6.1 A brief review

A typical control problem may be stated approximately as follows:

- Given a process to be controlled which may be ill-defined and little understood;
- Given desired control performances one would like to tolerate, which may be described by, for example, reference models, mathematical cost functions, performance index table, or fuzzy goals;
- Given a set of constraints imposed by characteristics of the physical system under consideration;
- Using available knowledge about the process and technologies construct a controller so that the overall control system behaves as would be expected in terms of satisfying and maintaining the prespecified performances.

In a general sense, the above control problem may be solved in various ways including classical feedback control, knowledge-based control, fuzzy control or neural network control, depending on a variety of factors. Conceptually, fuzzy-neural control is nothing more than a combined technology integrating some ideas and paradigms existing in the above approach, in particular those of fuzzy and neural control, to solve either a problem already solved using the other approach or new problems. However, it is useful to specify some apparent features of this approach so as to differentiate it from others. More specifically, by fuzzy-neural control, we normally mean that: (a) the controller itself has a structure resulting from the combination of fuzzy systems and neural networks; (b) the resulting control system consists of fuzzy systems and neural networks as independent components performing different tasks; or (c) the design methodologies for constructing respective controllers are hybrid ones coming from ideas in fuzzy and neural control.

The majority of work done in fuzzy-neural control can be described under the above meanings. For example, Yamaguchi *et al.* (1992) presented a fuzzy neural network called FAMOUS acting as a controller. IF-THEN like rules are stored in a *bidirectional associative memory*. FAMOUS can refine knowledge that is initially built-in by using neural network learning algorithms. Harris *et al.* (Brown and Harris 1992; Harris *et al.* 1992) made extensive comparisons between fuzzy logic control and neural network control. By drawing the equivalence between weights of neural networks and rule confidences, they developed three training algorithms for self-organizing fuzzy controllers. Kosko (Kosko 1992b; Kong and Kosko 1992) introduced a hybrid fuzzy-neural system named an *adaptive fuzzy system*. The structure of the system is built on fuzzy associative memory (FAM) rules with a reasoning algorithm similar to a data-driven forward chaining strategy. By viewing each FAM rule as defining a patch in the input–output product space, the FAM rules can be generated by using competitive clustering algorithms from a set of numerical data pairs. By extending Anderson's work (Anderson 1989) to include fuzzy control algorithms represented by a network structure into the action selection network, Berenji (1992) described a fuzzy logic control architecture with a reinforcement learning technique.

### 1.6.2 Ideas and paradigms

There exist a variety of possibilities for combining fuzzy systems with neural networks and fuzzy control with neural control. Treating a fuzzy system and a neural network as two different computational elements, they can be configured at a system level in a hierarchical manner. The overall task is divided hierarchically into levels, some of which are completed by neural networks and the others by fuzzy systems. For example, a fuzzy system may be employed as a supervisor at a higher level supervising a low-level neural network controller which directly controls the plant. They can also be configured in a parallel fashion, working together competitively or cooperatively. We will not pursue this kind of integration further in this book. In what follows, we will be concerned primarily with the combination of two paradigms at a component level, that is, the resulting fuzzy-neural system is regarded as integrated elements which can be used for various control purposes.

Basically, two points should be borne in mind when considering this integration: (a) What are the most useful characteristics of fuzzy systems and neural networks which are desirable to be present in the integrated system? (b) In what manner should these features be combined so as to enable the integrated system to be more generic and more powerful? We have outlined the first point in the previous section. The following discussions pertain to the second point.

Recall that three issues concerning the implementation of fuzzy control (knowledge representation, reasoning and acquisition) can be loosely expressed as *structure, computing and learning* in terms or neural networks. These correspondences not only provide a good starting point for our discussion which follows, but also give a clear viewpoint toward a better understanding and implementation of the integration. In this view, the solution to a control problem using fuzzy-neural approaches may be formulated as follows: given a plant $P$, a constraint set $Q$, and a performance $R$, choose a controller structure $S$ and a computing algorithm $T$; use prior knowledge $U$ and learning algorithms $V$ to construct $S$ so as to achieve $R$ subject to $Q$. More compactly, for a specific control problem $(P, Q, R)$, our task is to construct $(S, T)$ using $(U, V)$. Without reference to any detailed algorithms and implementations, we will restrict the following discussion to general considerations.

*A structure* $S$ is used to represent knowledge by its connection weights and units, and *computing* $T$ is concerned with operating stored knowledge in response to incoming input with interpolative capability. Normally, there exist certain dependencies between $S$ and $T$. To develop a fuzzy-neural structure, two ideas can be followed: the first is to neuralize existing fuzzy systems and the second one, vice versa, is to fuzzify existing neural networks.

- By *neuralizing fuzzy systems*, we mean to introduce neural concepts into fuzzy systems. Technically, it may be realized by mapping a fuzzy system into a neural network, either functionally or structurally, resulting in the neural network structures being either distributed or localized. In the former case, the input–output equivalence of both systems is of primary concern. Knowledge traditionally represented by IF-THEN rules is implicitly and distributively represented by connection weights and local processing units in the network. Each unit is responsible for many rules with different activation strengths. In other words, the function of each IF-THEN rule is undertaken by many units. It seems that a distributed network inherently possesses some fuzziness which may come from the distributiveness of the network although the term fuzziness here is used more technically and less conceptually. On the other hand, instead of paying primary attention to the functional aspect, the localized method is concerned mainly with the structural aspect, that is, to seek a close structural equivalence between the logic-based algorithm and the neural network structure. In this regard, each IF-THEN rule can be represented by only one unit in the network with associated weight vectors accounting for the IF part and THEN part. Thus, knowledge is represented locally in the sense that one piece of knowledge is embodied into one unit. However, the computing (reasoning) process for deriving an appropriate action in response to an input may be competitive or cooperative depending on the corresponding reasoning algorithms. Much more detail about these two approaches will be found in Chapters 6 and 9 respectively.
- *Fuzzifying neural networks* implies introducing fuzzy concepts into neural networks. As is well recognized, fuzzy systems are logic-based with fuzzy set representation and flexible fuzzy logic operations. Thus, the resulting fuzzy-neural system may include minimum, maximum, or even composition operators apart from usual sum and product operators found in neural computing. For example, a

product-sum unit may be replaced by a min-max unit. Chapter 10 will reconsider two well-known networks RBF and CMAC in the above sense. It should be pointed out that there is clearly no distinction between saying neuralizing and fuzzifying. What we describe here is in a large part for easy representation.

After the structure of a fuzzy-neural system is specified with a corresponding computing algorithm, the next concern might be what role will the system play in solving the control problem. Further, once the role is determined, the next issue would be how the fuzzy-neural system can be constructed and implemented. We refer to the latter matter as *learning* in a broad sense as discussed in Section 1.5.1. It is useful to highlight the following learning concepts relevant to the control application.

- *Active learning from an environment versus passive learning from samples:* Active learning from an environment without necessarily assuming the availability of training samples is much more demanding and difficult than passive learning from samples. Conventional supervised and unsupervised learning are examples of passive learning, whereas distal supervised learning (Jordan and Rumelhart 1992) is an example of active learning. In this case, the learner must be active in getting training samples from the environment. It is worth pointing out that supervised learning is closely related to parameter estimation in system identification and adaptive filtering, see, for example Chen and Billings (1992) and Nerrand *et al.* (1993) for more information.
- *On-line learning versus off-line learning:* On-line or real-time learning in control is not only normally desirable but also sometimes indispensable. Thus, the simplicity and efficiency of the learning algorithm are main concerns in order for it to be real-time. Notice that active learning is normally also on-line learning.
- *Dynamical variable-structure learning versus static fixed-structure learning:* Conventional neural networks training is fixed-structure learning. Structure parameters such as the number of units and layers remain unchanged during training. The learning implies a determination of connection weights only, with an implicit assumption that the present structure is capable of representing in some sense the desired functional relationship. In contrast, dynamical variable-structure learning is capable of self-determining some structure parameters such as the number of units in response to incoming data, thereby making the net structure dynamically variable. This feature is sometimes referred to as self-organizing for obvious reasons.
- *Global learning versus specific learning:* Depending on the usage, a fuzzy-neural system can be made purposely to be globally or specifically learning by supplying global or specific knowledge to the learner. If a fuzzy-neural system is used for a forward model, global learning is possible and normally necessary. In contrast, if it is employed as a direct controller it is unlikely that the controller will

be able to be global, capable of dealing with a variety of different situations which may be unforeseen and time-varying.

- *Spatial learning versus temporal learning:* A fuzzy neural controller may be constructed via spatial learning or temporal learning. A spatial learning scheme typically involves the extensive use of past experience to improve the present and future performance. The past experience can be cumulated during several past trials along a spatial direction in a time period of interest. Memorizing past control knowledge and repeatedly interacting with the controller environment are typical characteristics of spatial learning. Compared with spatial learning, temporal learning emphasizes learning accruing along a time direction only by using the information gained at immediate past time-instants and reacting to temporal variations. Learning control (Arimoto 1990) is a typical example of spatial learning, and adaptive control (Astrom and Wittenmark 1989) is a representative of temporal learning. In this book, the term learning control is reserved for reference to temporal learning only, unless otherwise specified.

The methodology for constructing a fuzzy-neural controller via learning and applying various learning concepts introduced above depends largely on the role of a fuzzy-neural system taking part in the control system. Like fuzzy systems and neural networks, a fuzzy-neural system can be incorporated into many well-established control architectures, acting as a forward or inverse model of the process being controlled. *Direct inverse control, internal model control* and various *predictive-based controls* are applicable configurations among others. Another possibility is to use fuzzy-neural systems, again as forward models, to cancel plant nonlinear functions in an affine nonlinear dynamical system, thereby enabling an input–output plant linearization to be approximately achieved. Essentially, the controller (in the form of a forward or inverse model) in these systems can be constructed via passive and on-line or off-line learning from samples which are assumed to be available.

When a fuzzy-neural system is used as a direct feedback controller similar to that of a fuzzy controller in Figure 1.3, a problem arises immediately as to where the training rules come from; the same problem as encountered in the design of algorithm-based fuzzy controllers. It is interesting to note that this problem is precisely equal to the question of where the teacher signals (training samples) come from as posed in deriving neural-controllers, and is extensively discussed by Barto (1990). Thus, the lack of experts who provide control rules in fuzzy systems is equal to the lack of teachers in neural networks who supply training samples. It is clear that, in this case, neither fuzzy systems nor neural networks nor fuzzy-neural systems, in their own rights, can provide immediate solutions, simply because the control rules and training samples are prerequisites for constructing these systems.

We have reviewed in Section 1.4 some solutions to deriving control rules in the context of fuzzy systems by using model-based or learning-based approaches. Likewise, considerable effort has been made to obtain teacher signals either explicitly or implicitly in neural control (Barto 1990; Narendra and Parthasarathy 1990). One of the methods is called *back-propagation through time* (Werbos 1990a, b; Nguyen and Widrow 1989; Wang and Miu 1990) or supervised learning with a distal teacher (Jordan and Rumelhart 1992). In this approach, the process model represented by another network is first identified, and then the tracking errors formed in the output of the process are back-propagated through the weight-fixed model network to the controller network. The other approach, referred to as *reinforcement learning* (Barto 1990; Barto *et al.* 1983; Anderson 1989), is to introduce a critic into the system to evaluate the control performance over a time period. The teacher signals or controller outputs are determined by considering whether they lead to increases in the plant performances specified by some meaningful measure. Further methods exist. For example, Ichikawa and Sawa (1992) proposed an approach using genetic algorithms to accommodate weights selection of a feedforward network-based controller. A genetic algorithm is used as a search strategy exploring directly the relationship between the controller parameters and the control performance, thereby avoiding an explicit need of teacher signals.

It is evident that constructing such a direct controller does represent a difficult and challenging issue, particularly when the process to be controlled is multivariable with strong interaction between variables. Active and on-line learning from the environment are essential. Also, dynamical variable-structure learning is desirable. A substantial portion of this book is devoted to exploring various possibilities of constructing direct controllers by learning. It is worth pointing out that although the introduction of a fuzzy-neural structure into a control configuration does not itself lead to the emergence of any novel learning paradigms in a radical way, this does offer some assistance in either speeding-up or improving the learning process. This can be done directly or indirectly, depending on the architecture of a fuzzy-neural system, incorporating any human control knowledge described by IF-THEN rules into the controller as, for example, initial weight parameter settings.

## 1.7 Objective and outline

The theme of this book is to try to provide a unified and comprehensive treatment of fuzzy and neural control systems under the divisions of knowledge representation, reasoning and acquisition which are recast technically into more practical issues, being appropriately termed as structure, computing, and learning for use in intelligent control. Our primary objective is to develop systems which are capable of performing self-organizing and self-learning functions in a real-time manner within multivariable system environments by utilizing fuzzy logic, neural networks, and a combination of both paradigms with emphasis on the novel system structures, algorithms, and applications to certain problems found in industrial and biomedical systems. Considerable effort is devoted to making the proposed approaches as generic, simple, and systematic as possible.

Throughout this book, it is assumed that the controlled processes are multi-input multi-output systems with possible time-delays in the control channels, no mathematical model for the process exists, but that some knowledge about the process is available. As an application, problems of multivariable control of blood pressure have been extensively studied with respect to each proposed structure and algorithm, aimed at demonstrating the utility and feasibility of the proposed approaches in solving relatively complex control problems, in particular those problems where neither control experts nor mathematical models of the controlled process are available.

The book essentially consists of two independent but closely related parts. The first part (Chapters 2-5) involves mainly the utilization of fuzzy algorithm-based schemes including topics such as approximate reasoning models, multivariable fuzzy controller structures, and learning algorithms. The second part (Chapters 6-10) deals primarily with the subjects of fuzzy-neural network-based approaches containing four hybrid fuzzy-neural controller structures with corresponding self-learning and self-organizing algorithms, covering five fundamental and widely used network paradigms including BNN, CPN, CMAC, RBF, and LVQ networks.

Figure 1.6 shows a schematic outline of the book. Starting from the same objectives, the book consists of two broad methodologies: fuzzy algorithm-based schemes and fuzzy-neural network-based approaches. Depending on the availability of the rule-base in the former case and the network structures of fuzzy-neural systems in the latter case, each of them is further divided into two subclasses as indicated in Figure 1.6. The dotted line signifies that the fuzzy-neural system-based approaches are dependent on the fuzzy algorithm-based schemes. The details of the following Chapters 2-11 are outlined as follows.

- Chapter 2: Taking account of the fuzzy nature of human decision making processes and real-time properties, Chapter 2 establishes a unified approximate reasoning model suitable for various definitions of IF-THEN, AND and ALSO linguistic connectives. Both fuzzy and random uncertainties can be coped with in the model.
- Chapter 3: By means of the unified approximate reasoning model derived in Chapter 2, Chapter 3 describes a viable decentralized fuzzy controller structure consisting of rule-based fuzzy controllers and a simple compensator unit . Extensive simulations on the problem of multivariable blood pressure control are presented.
- Chapter 4: This chapter presents a novel method capable of constructing rule-bases automatically via self-learning for use in fuzzy controllers. After introducing the system structure, the chapter focuses on the issues of the learning algorithm. Three learning update laws are suggested and the convergence property of the learning algorithms is analysed in the sense of some defined norms.
- Chapter 5: Continuing from Chapter 4, Chapter 5 is devoted to presenting a methodology for constructing the rule-base from learning data. By defining some performance measures, the behaviour of the proposed system, in terms of the learning ability (adaptability), reproducibility and robustness, is evaluated through

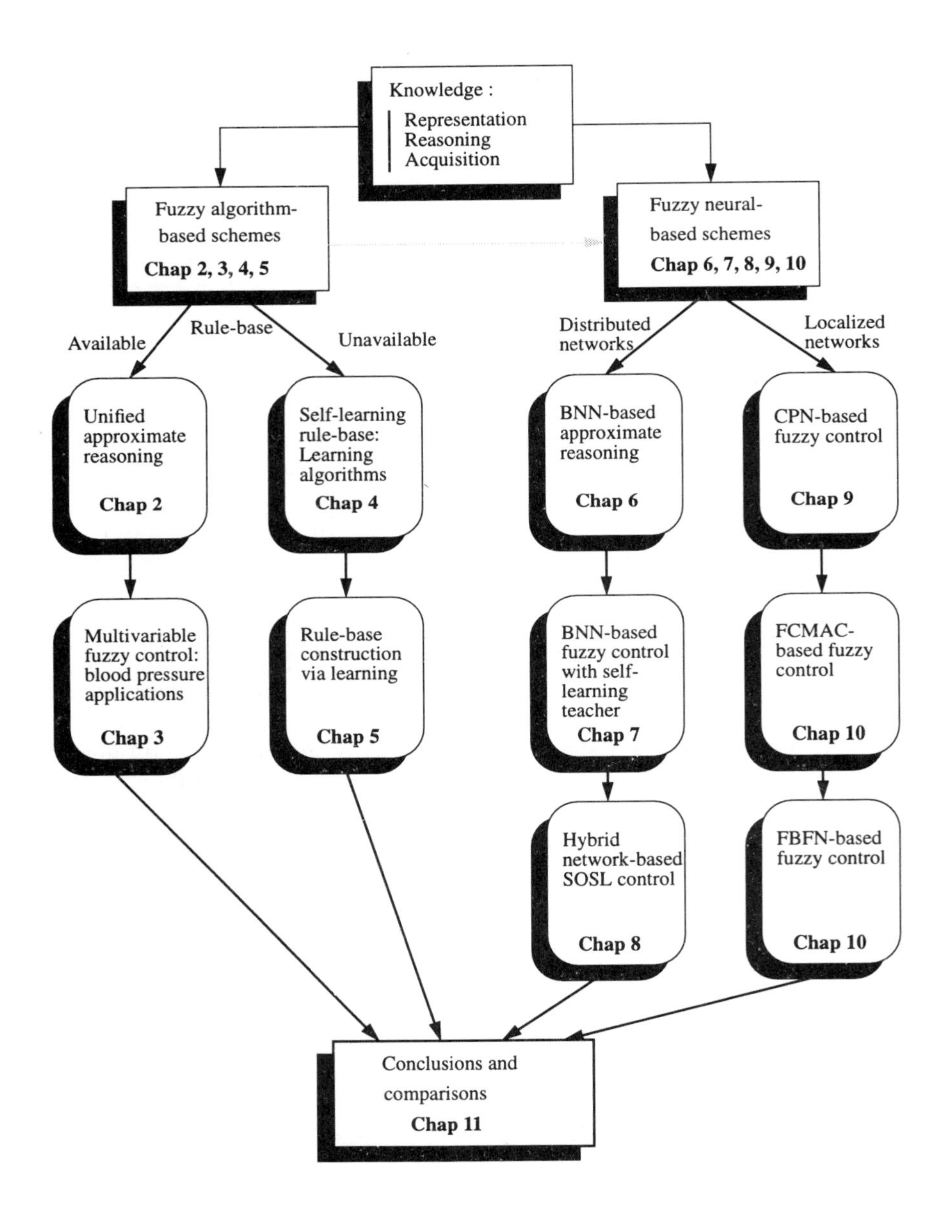


**Figure 1.6** A schematic outline of the book

a number of simulation studies on the blood pressure control problem.

- Chapter 6: By viewing the given rule-base as defining a global linguistic association constrained by fuzzy sets, Chapter 6, based on the principle of functional equivalence, demonstrates that the knowledge representation and approximate reasoning can be carried out by a Back-propagation Neural Network (BNN) with the aid of fuzzy set theory.
- Chapter 7: Continuing from the BNN-based reasoning strategy developed in Chapter 6 and making a crucial assumption that neither control experts nor teacher signals are available, Chapter 7 describes a multi-stage mechanism which is able to extract teacher signals in a systematic manner from the process being controlled. Three detailed structures for the controller implementation are discussed.
- Chapter 8: Still focusing on the topic of functional mapping, a hybrid fuzzy control system with the ability of self-organizing and self-learning is developed in Chapter 8 consisting of a variable-structure competitive network and a standard BNN network. The corresponding self-organizing and adaptive learning algorithms are given.
- Chapter 9: Shifting from a functional mapping to a structural mapping between fuzzy rule-based system and neural networks, Chapter 9 presents a simple but efficient scheme for structurally mapping a simplified fuzzy control algorithm into a counterpropagation network (CPN). The system is characterized by explicit representation, self-construction of rule-bases, and fast learning speed.
- Chapter 10 : This chapter presents two fuzzified networks FCMAC and FBFN by introducing some fuzzy concepts into well-known CMAC and RBF networks. The approach for constructing such kinds of fuzzy-neural controllers is given.
- Chapter 11: This concluding chapter closes the book by identifying key contributions of the book, comparing various learning paradigms developed in the book, and pointing out some future research directions relevant to fuzzy-neural control systems.

CHAPTER 2

# A unified approximate reasoning approach

*A unified approximate reasoning model is established suitable for various definitions of IF-THEN, AND and ALSO linguistic connectives. Both fuzzy uncertainties can be coped with in the model.*

## 2.1 Introduction

As far as expert control systems are concerned, two distinct characteristics can be noted. One is that the knowledge provided by experts or operators in a specific domain is usually qualitative and usually contains some kind of uncertainty. Therefore, it is inevitable that both knowledge representation and reasoning must be fuzzy or approximate. The other is that reasoning procedures must be completed within a predefined time interval. The first characteristic has resulted in the emergence of a theory known as Approximate Reasoning (AR) (Gupta *et al.* 1985) and the second characteristic requires that the computational models for representation and reasoning should be as simple as possible so that the reasoning calculations can be accomplished within a limited time. There exist some approximate reasoning schemes suitable for real-time applications. Among these algorithms, the one widely used is the relational matrix model, which was first adopted by Mamdani and is based on Zadeh's Compositional Rule of Inference (Zadeh 1973). Because rules can be interpreted as implication relations in a logical sense, numerous implication operators have been suggested in the literature (Mizumoto and Zimmermann 1982).

This chapter, taking account of the fuzzy nature of human decision-making processes and real-time properties, establishes a unified approximate reasoning model. It is mainly based on Possibility Theory developed by Zadeh (1978a) rather than on Relational Matrix Computation. The proposed algorithms are suitable for various definitions of IF-THEN, AND and ALSO linguistic connectives frequently used in rule-based expert systems. Some existing reasoning methods can be regarded as special cases of this unified approach. Both fuzzy and random uncertainties can be incorporated into the model. The proposed computational algorithms may provide another possibility for on-line reasoning in real-time expert systems.

## 2.2 Reasoning models

### 2.2.1 Statement of the problem

A rule in a typical rule-based expert system is usually expressed in the form of an IF-THEN statement. There exist a variety of terms which are used to describe such a rule. Examples are condition/action, precedent/consequent, premise/conclusion and input–output (primarily used in expert control). This book will mainly use the terms condition/action and input–output.

Assume that the system we are interested in has $n$ inputs and a single output and that these are denoted by $X_1, X_2, \cdots, X_n$ and $U$. Furthermore, it is assumed that a set of $m$ linguistic rules, which are represented in the form of "IF <CONDITION> THEN <ACTION>", have been obtained from domain expert(s). Then the problem we are concerned with may be described as follows:

- Given rules: $R^1$ ALSO $R^2$ ALSO $\cdots$ ALSO $R^m$ where $R^j$ has the form:
  IF $X_1$ is $A_1^j$ AND $X_2$ is $A_2^j$ AND $\cdots$ AND $X_n$ is $A_n^j$
  THEN $Y$ is $B^j$, $j = 1, 2, \cdots, m$.
- Given input data: $X_1$ is $C_1$ AND $X_2$ is $C_2$ AND $\cdots$ AND $X_n$ is $C_n$
- To find output data: $Y$ is $D$, $D$ = ? where $X_1, X_2, \cdots, X_n, Y$ are variables whose values are taken from the universes of discourse $U_1, U_2, \cdots, U_n$ and $V$ respectively. $A_i^j$, $C_i$, $B^j$, $D$ are fuzzy subsets which are defined on the corresponding universe, and represent some fuzzy concepts such as *big, medium* and *small* etc.

It is worth noting that for multi-input/multi-output systems it is easy to decompose each multi-output rule into several rules with only one output for each rule. For instance, suppose that there is a rule denoted as

IF $X_1$ is $A_1^j$ AND $X_2$ is $A_2^j$ AND $\cdots$ AND $X_n$ is $A_n^j$

THEN $Y_1$ is $B_1^j$ AND $Y_2$ is $B_2^j$ $j = 1, 2, \cdots, m$.

It can be treated as two different rules:

1. IF $X_1$ is $A_1^j$ AND $X_2$ is $A_2^j$ AND $\cdots$ AND $X_n$ is $A_n^j$ THEN $Y_1$ is $B_1^j$
2. IF $X_1$ is $A_1^j$ AND $X_2$ is $A_2^j$ AND $\cdots$ AND $X_n$ is $A_n^j$ THEN $Y_2$ is $B_2^j$

Similarly, rules like

IF $X_1$ is $A_1^j$ OR $X_2$ is $A_2^j$ OR $\cdots$ OR $X_n$ is $A_n^j$ THEN $Y$ is $B^j$
can be decomposed into $n$ rules with a form

IF $X_i$ is $A_i^j$ THEN $Y$ is $B^j$, $\qquad i = 1, 2, 3, \ldots, n$

Therefore, it is sufficient here only to discuss rules with one output.

In view of the problem stated above, there are basically two issues which must be solved. One is how to represent or translate the rules and observed data expressed in natural language form, for instance, $X_i$ is $A_i^j$, into computational language, and how to define the linguistic connectives such as IF-THEN, AND and ALSO etc. The other is how to infer the outputs, possibly in a real-time manner, provided that the rules and observed data are given, and at the same time ensure that the translation and inference are computationally simple. These two issues are called computational knowledge representation and reasoning in this thesis. It is obvious that the two issues are interrelated. It should be pointed out that the inference system must be able to reach a conclusion in the situation where none of the conditions for the $m$ rules are completely identical to the observed data. This ability is one of the most distinct features in approximate reasoning, and is notably different from classical reasoning methods such as *modus ponens*. Possibility theory (Zadeh 1978a) provides a framework to deal with these problems. The basic concept of the theory is described next.

### 2.2.2 Possibility distribution of data and rules

Possibility theory was developed by Zadeh as a complement to probability theory. It is intended to capture linguistic vagueness and nonstatistical uncertainty which exist commonly in human knowledge. The basic concept of possibility may be stated as follows. If $X$ is a variable taking values on a universe of discourse $U$, then the possibility distribution of $X$, denoted $\Pi_X$, is the fuzzy set of all possible values of $X$. More specifically, let $\pi_X(u)$ be the possibility that $X$ can take the value $u$, $u \in U$. Then the membership function of $X$ is numerically equal to the possibility function $\pi_X(u)$: $U \to [0,1]$, which associates each element $u \in U$ with the possibility that $X$ may take $u$ as its value. It is worth noting, as pointed out by Zadeh (1978b), that there is a significant difference between possibility and probability. Possibility relates to our perception of the degree of the feasibility or ease of attainment, whereas probability is largely associated with the degree of likelihood, belief, frequency or proportion. The importance of possibility lies in the fact that much of human decision-making is based on information which is possibilistic rather than probabilistic in nature. In addition, there is a distinction between the concepts of a possibility distribution function of a variable $X$ and that of a membership function of a fuzzy set of all possible values of $X$, although they are numerically equivalent to each other. The former measures the possibility of a variable taking a certain value in the specified universe of discourse, whereas the latter expresses the degree to which a certain element in the universe belongs to a fuzzy concept.

Now we focus our attention on the representation of data and rules. From the rule segment presented in Section 2.1, it can be seen that all the rules and observed data consist of a set of "$X$ is $F$" statements and linguistic connectives IF-THEN, AND and ALSO. The form "$X$ is $F$ "is called an atomic element, or a canonical form, which is composed of a variable $X$ and a value $F$.

1. "$X$ is $A$" induces a possibility distribution function (PDF) $\pi_X(u)$. More specifically, let $X$ be a variable which takes values on a universe of discourse $U$, and let $A$ be a fuzzy subset of $U$. For instance, the statement "Temperature is high" may be denoted as "$T$(degrees celsius) is $H$", where $T$ is a variable taking values in a specific temperature interval, say [20,100], and $H$ is a fuzzy set over that interval representing the fuzzy concept *high*. The statement "$X$ is $A$", according to possibility theory, induces a PDF $\pi_X(u)$. In the absence of any information regarding X other than that conveyed by the statement "$X$ is $A$", $\pi_X(u)$ is equal to $A(u)$, that is, $\pi_X(u) = A(u)$, $u \in U$, where $A(u)$: $U \rightarrow [0,1]$ is the membership function of $A$.
2. The modified atomic statement "$X$ is $h$ A" induces a modified PDF $\pi_X^*(u)$. Sometimes, linguistic hedges such as "not", "very" and "more or less" may be included in the atomic statement. An example is the statement "Temperature is very high". In this case, a modifier $h$ is used to alter the PDF $\pi_X(u)$ previously derived from "$X$ is $A$". Some commonly used hedges and corresponding definitions are given below.

$$\begin{aligned} h &= \text{"NOT"} && \pi_X^*(u) = A^*(u) = \textstyle\sum_u [1-A(u)]/u \\ h &= \text{"VERY"} && \pi_X^*(u) = A^*(u) = \textstyle\sum_u A^2(u)/u \\ h &= \text{"MORE OR LESS"} && \pi_X^*(u) = A^*(u) = \textstyle\sum_u A^{0.5}(u)/u \end{aligned} \tag{2.1}$$

   where $\sum$ denotes the union of singletons.
3. It is usually the case that rules and data consist of several atomic statements which are linked by the connective AND. Consider a compound statement "$X_1$ is $A_1$ AND $X_2$ is $A_2$ AND . . . AND $X_n$ is $A_n$", where $X_i$, $i = 1, 2, \ldots, n$ is the ith variable taking values on the universe $U_i$, and $A_i$, $i = 1, 2, \ldots, n$, is a fuzzy set defined over $U_i$. Then the statement induces a joint PDF $\pi_X(u_1, u_2, \ldots, u_n)$ over a space $U_1 \times U_2 \times \ldots \times U_n$:

$$\pi_X(u_1, u_2, \ldots, u_n) = \mathop{\Xi}_{i=0}^{n} \pi_{X_i}(u_i) \tag{2.2}$$

where $\pi_{X_i}(u_i) = A_i(u_i)$, and "$\Xi$" represents the connective AND. The computational definition of $\Xi$ will be discussed later.

4. Rule "IF $X$ is $A$ THEN $Y$ is $B$" induces a conditional PDF. In a rule-based expert system, the relationship between the condition(input) part and the action(output) part is usually presented in the form stated above. It should be noted that the rule can also be considered as an implication, meaning that "$X$ is $A$" implies "$Y$ is $B$", if it is viewed from a classical logic point of view. More specifically, let $X$ and $Y$ be variables taking values on the universes of discourse $U$ and $V$ respectively, and $A$ and $B$ be fuzzy subsets defined over $U$ and $V$, where $A(u)$: $U \rightarrow [0,1]$ and $B(v)$:$V \rightarrow [0,1]$. Then the rule "IF $X$ is $A$ THEN $Y$ is $B$" induces a conditional PDF over $U \times V$ given by

$$\pi_{Y|X}(u,v) = \pi_X(u) \;\; \Phi \;\; \pi_Y(v) \tag{2.3}$$

   where "$\Phi$" stands for the IF-THEN operation which will be defined subsequently.

5. Compound rules will generate a joint conditional PDF.
   Consider a rule

   *IF* $X_1$ *is* $A_1$ *AND* $X_2$ *is* $A_2$ *AND. . .AND* $X_n$ *is* $A_n$ *THEN* $Y$ *is* $B$

   where $X_1, X_2, \ldots, X_n$ and Y are variables taking values in $U_1, U_2, \ldots, U_n$ and V respectively; $A_1, A_2, \ldots, A_n$ and B are fuzzy sets defined on the corresponding universe. Then the rule induces a joint conditional PDF over the space $U_1 \times U_2 \times \ldots \times U_n \times V$ given by

$$\pi_{Y|X_1, X_2, \ldots, X_n}(u_1, u_2, \ldots, u_n, v) = \pi_X(u_1, u_2, \ldots, u_n) \;\; \Phi \;\; \pi_Y(v) \tag{2.4}$$

   where $\pi_X(u_1, u_2, \ldots, u_n) = \underset{i=0}{\overset{n}{\Xi}} \pi_{X_i}(u_i) \qquad u_i \in U_i \quad i = 1, 2, \ldots, n$

   Thus, both observed data and rules are represented in a unified approach under the framework of possibility theory. Next, the reasoning procedures will be presented in detail.

### 2.2.3 Reasoning model

As stated in Section 2.1, apart from the computational representation the other important issue involved in reasoning is the inference scheme, that is, how a reasonable conclusion should be derived in the light of rules and observed data.

Assume that both the rules and data are given in the following way:

- Rule $R^j$:

  IF $X_1$ is $A_1^j$ AND $X_2$ is $A_2^j$ AND . . . AND $X_n$ is $A_n^j$ THEN Y is $B^j$

- Input data:

  $X_1$ is $C_1$ AND $C_2$ is $A_2$ AND . . . AND $X_n$ is $C_n$

Then the present output data "$Y$ is $D$" based on the rule and input data can be inferred from the following procedures:

- First, observed data induces a joint PDF over $U_1 \times U_2 \times \ldots \times U_n$

$$\pi^*_{X_1, X_2, \ldots, X_n}(u_1, u_2, \ldots, u_n) = \Xi_{i=0}^{n} \pi^*_{X_i}(u_i) \qquad u_i \in U_i \tag{2.5}$$

  where $\pi^*_{X_i}(u_i)$ is a PDF created by the statement "$X$ is $C_i$". $\pi^*_{X_i}(u_i)$ is numerically equal to $C_i(u_i)$, the membership function of fuzzy set $C_i$. Consequently

$$\pi^*_{X_1, X_2, \ldots, X_n}(u_1, u_2, \ldots, u_n) = \Xi_{i=0}^{n} C_i(u_i) \tag{2.6}$$

- Second, rule $R^j$ generates a joint conditional PDF over $U_1 \times U_2 \times \ldots \times U_n \times$ V

$$\pi^j_{Y|X_1,X_2,\cdots,X_n}(u_1,u_2,\cdots,u_n,v) = \pi^j_{X_1,X_2,\cdots,X_n}(u_1,u_2,\cdots,u_n) \;\Phi\; \pi^j_Y(v) \tag{2.7}$$

  where "$\Phi$" denotes the IF-THEN operator, and $\pi^j_{X_1, X_2, \ldots, X_n}(u_1, u_2, \ldots, u_n)$ is induced by the *condition* part of $R_j$, via

$$\pi^j_{X_1, X_2, \ldots, X_n}(u_1, u_2, \ldots, u_n) = \Xi_{i=1}^{n} \pi_{X_i^j}(u_i) = \Xi_{i=1}^{n} A_i^j(u_i) \qquad u_i \in U_i$$

  and $\pi^j_Y(v)$ is induced by the *action* part of $R^j$, via $\pi^j_Y(v) = B^j(v) \qquad v \in V$

  Let $F^j(u) = \Xi_{i=1}^{n} A_i^j(u_i)$

  then,

$$\pi^j_{Y|X_1, X_2, \ldots, X_n}(u_1, u_2, \ldots, u_n, v) = F^j(u) \;\Phi\; B^j(v) \tag{2.8}$$

- Third, according to input PDF (2.5) and rule PDF (2.7), the output PDF $\pi_Y^{*j}(v)$ induced by "$U$ is $D$" can be obtained as follows:

$$\pi_Y^{*j}(v) = \pi^*_{X_1,X_2,\cdots,X_n}(u_1,u_2,\cdots,u_n) \;\mathrm{o}\; \pi^j_{Y|X_1,X_2,\cdots,X_n}(u_1,u_2,\cdots,u_n,v) \tag{2.9}$$

  where symbol "o" denotes the compositional inference operator. Substituting equations (2.6) and (2.8) into equation (2.9) and writing $\pi_Y^{*j}$ as $D^j(v)$, we obtain

$$D^j(v) = \left[\Xi_{i=1}^{n} C_i(u_i)\right] \mathrm{o} \left[F^j(u) \;\Phi\; B^j(v)\right] = \left[\Xi_{i=1}^{n} C_i(u_i)\right] \mathrm{o} \left\{\left[\Xi_{i=1}^{n} A_i^j(u_i)\right] \;\Phi\; B^j(v)\right\} \tag{2.10}$$

Because the main basis on which we are reasoning is how far or how near the present input data approaches the *condition* part of $R^j$, measurement of this approaching degree should be irrelevant to the *action* part of $R^j$. Consequently, it is reasonable to assume that the compositional operation of input data with respect to

$\left[F^j(u)\ \Phi\ B^j(v)\right]$ only has effect on the *condition* part of $R^j$, i.e. $F^j(u)$, regardless of the computational definition of "$\Phi$" . Thus,

$$D^j(v) = \left[\mathop{\Xi}_{i=1}^{n} C_i(u_i)\ \mathrm{o}\ \mathop{\Xi}_{i=1}^{n} A_i^j(u_i)\right]\ \Phi\ B^j(v) \tag{2.11}$$

Note that $C_i(u_i)$ and $A_i^j(u_i)$ denote the $i$th element (fuzzy subset) of the input data and of the rule $R^j$ respectively. Therefore, for a specific $i$, the only candidate among $A_i^j(u_i)$, $i = 1, 2, \ldots, n$ with which input data $C_i(u_i)$ interact is $A_i^j(u_i)$. Taking this fact into account, we get

$$D^j(v) = \left\{\mathop{\Xi}_{i=1}^{n}\left[C_i(u_i)\ \mathrm{o}\ A_i^j(u_i)\right]\right\}\ \Phi\ B^j(v) \tag{2.12}$$

If the Sup-Min compositional operator is used, then

$$C_i(u_i)\ \mathrm{o}\ A_i^j(u_i) = \mathop{Sup}_{u_i \in U_i} Min\left[C_i(u_i), A_i^j(u_i)\right] \tag{2.13}$$

Equation (2.13) is, in fact, a measurement of approaching or matching degree for two fuzzy subsets $C_i$ and $A_i^j$. It is called the Possibility Measure and is denoted by

$$\beta_i^j = Poss(C_i | A_i^j) = \mathop{Sup}_{u_i \in U_i} Min\left[C_i(u_i), A_i^j(u_i)\right] \tag{2.14}$$

$\beta_i^j$ has the following properties which illustrate further that it is a good similarity measurement. Assume that $C_i$ and $A_i^j$ are normalized fuzzy subsets, i.e. there is at least one member in the corresponding universe with a membership of 1. Then, when $C_i \subseteq A_i^j$, i.e. for all $k$, $C_i(u_{ik}) \le A_i^j(u_{ik})$, where $u_{ik} \in U_i$, then $\beta_i^j = 1$. When $C_i \cap A_i^j = \varnothing$, where $\varnothing$ denotes the null set, i.e. for all $k$, $\min\left[C_i(u_{ik}), A_i^j(u_{ik})\right] = 0$, $u_{ik} \in U_i$, then $\beta_i^j = 0$. Otherwise $0 < \beta_i^j < 1$, which means two atomic elements in the input data and in the rule have a similarity measured by $\beta_i^j$.

Thus, we finally obtain

$$D^j(v) = \left[\mathop{\Xi}_{i=1}^{n} \beta_i^j\right]\ \Phi\ B^j(v) = \beta^j\ \Phi\ B^j(v) \tag{2.15}$$

The above equation is the approximate reasoning computational model under a single rule. It is worth noting that "$\Xi$" and "$\Phi$" stand for the linguistic connectives AND and IF-THEN. Therefore it is very convenient to obtain different reasoning algorithms if "$\Xi$" and "$\Phi$" are described using different computational definitions. The model can be interpreted as follows. First the Possibility Measure $\beta_i^j$ for every $i$ is calculated. Then a global measure $\beta^j$ is derived from the $\beta_i^j$s with a predefined AND ($\Xi$) computation. Finally, the action part $B^j(v)$ is modified to $D^j(v)$ according to $\beta^j$ and the predefined IF-THEN ($\Phi$) operation.

## 2.3 Rule aggregation and operator selection

### 2.3.1 Rule aggregation

There are usually many rules in the rule-base of practical expert systems. In classical rule-based systems, a rule is fired if input data exactly match the rule's *condition* part so that the final decision made would be the *action* part of the fired rule. However, in fuzzy rule-based systems, because a rule's *condition* part is involved in the fuzzy predicates, it is impossible to ensure that for every possible input data there always exists one rule which is completely identical to that data. As a consequence, it is usually the case that more than one rule will make contributions to a final decision. This fact gives rise to an issue which may be called rule aggregation. In the majority of early fuzzy reasoning systems, especially in fuzzy control, $m$ rules were first represented by the $m$ relational matrices with the same dimension for each matrix. A final reasoning matrix was then constructed by means of a maximum operation on corresponding elements of $m$ matrices. In other words, the aggregation took place before the input data entered the system. In contrast, an aggregation method is proposed here which is carried out after every rule has been evaluated with respect to the current input data.

1. Rule firing: Referring to equation (2.13), the Possibility Measure $\beta^j$ is a good indicator which signifies the similarity degree to which input data match a rule $R^j$. The more similar they are, the more close to 1 is $\beta^j$. It will be desirable to fire only those rules whose possibility measures $\beta^j$ are greater than a prespecified threshold $\beta_0$, where $0 \leq \beta_0 \leq 1$. Therefore, only these fired rules will make contributions to the final action.
   An alternative way to determine if a rule should be fired is proposed by employing the concept of *sigma-count*, discussed by Zadeh (1989). A *sigma-count* of a fuzzy set $F$ defined on a universe of discourse $Z$ is defined as the arithmetic sum of the grade of membership of $F$, ie

$$\sum count(F) = \sum_{i=1}^{L} F(z_i) \qquad z_i \in Z \tag{2.16}$$

   where $L$ is the element number in the support of $F$. The *sigma-count* is basically an analogue to the cardinality of ordinary sets. The relative *sigma-count*,denoted by $\sum$count($F/G$), is defined as

$$\sum count(\frac{F}{G}) = \frac{\sum count(F \cap G)}{\sum count(G)} \tag{2.17}$$

   where $F \cap G$, the intersection of $F$ and $G$, is defined by $F \cap G(z_i) =$ Min[$F(z_i)$, $G(z_i)$]. Thus,

$$\sum count\left(\frac{F}{G}\right) = \frac{\sum_{i=1}^{L} Min\,[F(z_i), G(z_i)]}{\sum_{i=1}^{L} G(z_i)} \tag{2.18}$$

The relative *sigma-count*, in fact, expresses the proportion of the element $F$ which is in $G$.

In order to use this concept, it is necessary to build two fuzzy sets. Assume that for the current input data and a rule $R^j$, $n$ $\beta_i^j$s are obtained according to equation (2.14), and that for all $i$ $\beta_i^j > 0$. Then a fuzzy set T can be constructed on a universe of discourse $\{1, 2, \ldots, n\}$ such that $T(i) = \sum_i B_i^j/\text{i}$, where $\sum_i$ stands for the union of singletons, and $i = 1, 2, \ldots, n$. Likewise, an ordinary set $W$ is defined as $W(i) = \sum_i 1/i$, $i = 1, 2, \ldots, n$, with the understanding that every member of $W$ has a membership of 1. By means of the definition of relative *sigma-count*, denoted as $\tau^j$, we obtain

$$\tau^j = \sum count\left(\frac{T}{W}\right) = \frac{\sum_i^n Min\,[T(i), W(i)]}{\sum_i^n W(i)} \tag{2.19}$$

By substituting for $T(i)$ and $W(i)$ defined above in equation (2.19), $\tau^j$ can be written as

$$\tau^j = \frac{\sum_{i=1}^{n} \beta_i^j}{n} \tag{2.20}$$

$\tau^j$ can be considered as a measurement of the proportion of matched atomic elements (with respect to input data) which are in $n$ atomic elements of the *condition* part of rule $R^j$. If every element is completely matched, then $\tau^j = 1$. This measurement may be used to deal with a linguistic statement such as "if *most* of the rule's conditions are satisfied then the rule is fired", or "if *at least half* of the rule's conditions are satisfied then the rule is fired". These are treated by Yager (1984) as qualified rules. Again, a threshold $\tau_0$ is needed to determine whether or not a rule should be considered as a fired rule, by comparing $\tau^j$ with $\tau_0$.

2. Rule combination: As stated previously, more than one rule tend to contribute to the final decision, but with different strengths indicated by $\beta^j$. Thus, a mechanism which combines or aggregates all contributed rules is needed. It should be stressed that some assumptions must be made before a link mechanism is built. These assumptions should include: all rules in the rule base must be consistent, meaning that no conflict rule exists; the rule-base should be complete, implying that for every possible input data there always exists at least one rule $R^j$ for

which $\beta^j > 0$; and each rule is independent from others in the sense that there is no causal relationship between the rules.

Let us assume that there are $m$ rules in the rule-base . Each rule $R^j$ has the form

$$IF\ X_1\ is\ A_1^j\ AND\ X_2\ is\ A_2^j\ AND\ \ldots\ AND\ X_n\ is\ A_n^j\ THEN\ \ Y\ is\ B^j.$$

Also, $m$ rules are connected by ALSO as

$$R^1\ \ ALSO\ \ R^1\ \ ALSO\ldots ALSO\ \ R^m;$$

When the current input data

$$X_1\ \ is\ \ C_1\ \ AND\ \ X_2\ \ is\ \ C_2\ \ AND\ \ \ldots\ \ AND\ \ X_n\ \ is\ \ C_n$$

are given, then the current output data can be derived according to the following procedures.

Suppose that there are $K$ rules being fired, which are determined either by $\beta^k \geq \beta_0$, $k \in [1,K]$ or by the relative *sigma-count* $\tau \geq \tau_0$. Referring to equation (2.15), the $k$th output data "$Y$ is $D^k$" obtained from the $k$th rule $R^k$ are

$$D^k(v) = \left[ \mathop{\Xi}_{i=0}^{n} \beta_i^k \right] \Phi\ B^k(v) \tag{2.21}$$

Then the global output data "$Y$ is $D$" induces a PDF

$$\pi_Y(v) = \mathop{\Theta}_{k=1}^{K} \pi_Y^{*k}(v) = \mathop{\Theta}_{k=1}^{K} D^k(v) \tag{2.22}$$

More specifically

$$\pi_Y(v) = \mathop{\Theta}_{k=1}^{K} \left\{ \left[ \mathop{\Xi}_{i=0}^{n} \beta_i^k \right] \Phi\ B^k(v) \right\} \tag{2.23}$$

where "$\Theta$" stands for the linguistic connective ALSO.

### 2.3.2 Operator selection and some selected algorithms

A variety of reasoning algorithms can be obtained easily by giving the computational definitions of the ALSO operation "$\Theta$", the AND operation "$\Xi$" and the IF-THEN operation "$\Phi$" in equation (2.23). It is obvious that employing different definitions will generally give different results. It has been shown that operator selection is of significant importance in determining the performance of an expert system. Unfortunately there are no general criteria guiding the selection because it is very difficult to evaluate the performance of different operator definitions from a theoretical point of view. It is widely recognized that the selection is strongly application-specific. However, there exist some definitions which can be demonstrated practically to be better selections.

The aggregating connective ALSO is usually interpreted as a "UNION" operation denoted by $\cup$ which corresponds to the maximum operator. It can also be interpreted as an "INTERSECTION" operation denoted by $\cap$ corresponding to the minimum operator denoted by "$\Lambda$". This is particularly true if Lukasiewicz logic implication operators discussed below are adopted. The AND operation can be defined as either a minimum or algebraic product operator denoted by "*". Numerous definitions exist for the IF-THEN operation (Mizumoto 1987). Some of them are defined by generalizing classical logic implication operators for a fuzzy situation, whereas others are defined purely from practical considerations. Four widely used definitions are given here as follows.

Suppose that the rule is of the form "IF $X$ is $A$ THEN $Y$ is $B$". Then the conditional PDF induced by the rule is given by

$$\pi_{Y|X}(u,v) = \pi_X(u)\ \Phi\ \pi_Y(v) = A(u)\ \Phi\ B(v)$$

Various IF-THEN ($\Phi$) operators have been defined by

Mamdani ($\Lambda$): $\pi_{Y|X} = A(u)\Lambda B(v)$

Larsen (*): $\pi_{Y|X} = A(u)*B(v)$

Zadeh (Z): $\pi_{Y|X} = 1\Lambda[1-A(u) + B(v)]$

Godel (G): $\pi_{Y|X} = \begin{cases} 1 & \text{if } A(u) \le B(v) \\ B(v) & \text{if } A(u) > B(v) \end{cases}$

By combining three operators, expressing them as a triplet ( $\Theta$, $\Xi$, $\Phi$ ) and substituting them into equation (2.23), various reasoning algorithms can be derived easily, some of them being presented below:

1. $(\cup, \Lambda, \Lambda)$: $$D(v) = \bigcup_{k=1}^{K}\left[\left(\mathop{\Lambda}_{i=1}^{n} \beta_i^k\right) \Lambda\ B^k(v)\right] \qquad (2.24a)$$

2. $(\cup, \Lambda, *)$: $$D(v) = \bigcup_{k=1}^{K}\left[\left(\mathop{\Lambda}_{i=1}^{n} \beta_i^k\right) *\ B^k(v)\right] \qquad (2.24b)$$

3. $(\cup, *, \Lambda)$: $$D(v) = \bigcup_{k=1}^{K}\left[\left(\mathop{*}_{i=1}^{n} \beta_i^k\right) \Lambda\ B^k(v)\right] \qquad (2.24c)$$

4. $(\cup, *, *)$: $$D(v) = \bigcup_{k=1}^{K}\left[\left(\mathop{*}_{i=1}^{n} \beta_i^k\right) *\ B^k(v)\right] \qquad (2.24d)$$

5. $(\cap, \Lambda, Z)$: $$D(v) = \bigcap_{k=1}^{K}\left\{1\ \Lambda\left[1-\left(\mathop{\Lambda}_{i=1}^{n} \beta_i^k\right) + B^k(v)\right]\right\} \qquad (2.24e)$$

6. $(\cap, *, Z)$: $$D(v) = \bigcap_{k=1}^{K}\left\{1\ \Lambda\left[1-\left(\mathop{*}_{i=1}^{n} \beta_i^k\right) + B^k(v)\right]\right\} \qquad (2.24f)$$

7. $(\cap, \wedge, G):\quad D(v) = \bigcap_{k=1}^{K} \begin{cases} 1 & \text{if } [\bigwedge_{i=1}^{n} \beta_i^k] \leq B^j(v) \\ B^k(v) & \text{if } [\bigwedge_{i=1}^{n} \beta_i^k] > B^k(v) \end{cases}$ (2.24g)

8. $(\cap, *, G):\quad D(v) = \bigcap_{k=1}^{K} \begin{cases} 1 & \text{if } [\mathop{*}_{i=1}^{n} \beta_i^k] \leq B^j(v) \\ B^k(v) & \text{if } [\mathop{*}_{i=1}^{n} \beta_i^k] > B^k(v) \end{cases}$ (2.24h)

## 2.4 Reasoning with uncertain data and rules

By asserting that much of human decision-making is based on possibilistic rather than probabilistic information, we mean that real-world phenomena are first perceived and fused by human cognition mechanisms and then expressed in linguistic form. This kind of information is qualitative and non-numerical in nature. It is this perception, fusion and expression which makes the information we are dealing with fuzzy, and it is the ability of fuzzy set theory and possibility theory to handle the fuzzy information in numerical form that makes the theories very useful in expert systems.

On the other hand, it should be noted that the fuzzy information provided by humans in expert systems may not be certain in the sense that it is not completely sure that the value of a variable $X$ belongs to the support of $A$. The uncertainty may be present in both input data and rules. For example, the statement "$X$ is $A$ with certainty $\mu$" signifies that the certainty with which the variable $X$ takes the value $u$ in $A$ with possibility equal to $A(u)$ is $\mu$. At the same time, the certainty with which $X$ takes the value $u$ in all fuzzy subsets (except $A$) of power set of $U$ should be 1-$\mu$.

### 2.4.1 Uncertainty with data

Assume that the current input data are of the form

$$X_1 \text{ is } C_1[\mu_1] \text{ AND } X_2 \text{ is } C_2[\mu_2] \text{ AND } \ldots \text{ AND } X_n \text{ is } C_n[\mu_n] \quad (2.25)$$

where $\mu_i \in [0,1]$, and "$X_i$ *is* $C_i[\mu_i]$" should be understood as "$X_i$ is $C_i$ with certainty $\mu_i$". Yager (1984) has suggested a mechanism for representing this kind of statement. More specifically, if $E$ and $F$ are fuzzy subsets of $U$, then "$X_i$ is $E_i$ with certainty $\mu_i$" can be modified to "$X_i$ is $F_i$" , where for any $u \in U$

$$F_i(u) = Min\ [\mu_i\ ,\ E_i(u)] + (1-\mu_i) \tag{2.26}$$

If $\mu_i = 1$, i.e. complete certainty, then $F_i = E_i$. In this case, all the previous results are applicable. On the other hand, if $\mu_i = 0$, i.e. complete uncertainty, then $F_i = U$ means that there is no information regarding the value of $X$. Thus, input data become the following, where a certainty of 1 is implied

$$X_1\ is\ \overline{C}_1\ AND\ X_2\ is\ \overline{C}_2\ AND\ \ldots\ AND\ X_n\ is\ \overline{C}_n$$

where $\overline{C}_i(u)$ is defined by equation (2.26). It can be proved easily that $\overline{C}_i(u)$ is still a normal set, so that the reasoning model discussed previously can be applied to infer the output data with this modified input.

Instead of modifying the input data, another approach is proposed here in which the possibility measure is modified. Suppose that the input data have the same form as equation (2.25) but with $\mu_i \in [0,1]$. The reason for excluding the situation of $\mu_i = 0$ is that reasoning with such completely uncertain data is meaningless and therefore it is assumed that the data are more or less certain, i.e. $\mu_i \in [0,1]$. Furthermore, suppose that $n$ possibility measures $\beta_i^j$, $i = 1, 2, \ldots, n$ are derived such that

$$\beta_i^j = Poss(\ C_i\ |\ A_i^j\ ) \qquad i = 1, 2, \ldots, n$$

where uncertainty is not taken into account. Then the modified possibility measure is given by

$$\overline{\beta}_{ai}^j = Min\ [\beta_i^j\ ,\ \mu_i] \tag{2.27a}$$

or by

$$\overline{\beta}_{bi}^j = \beta_i^j * \mu_i \tag{2.27b}$$

If $\mu_i = 1$ then $\overline{\beta}_{ai}^j = \overline{\beta}_{bi}^j = \beta_i^j$; otherwise the similarity degree indicated by the possibility measure is reduced from $\beta_i^j$ to $\overline{\beta}_{ai}^j$ or to $\beta_{bi}^j$. $\overline{\beta}_{bi}^j$ is more conservative than $\overline{\beta}_{ai}^j$ in the sense that $\mu_i$ is always taken into account in $\overline{\beta}_{bi}^j$, whereas $\mu_i$ is sometimes neglected when the certainty is higher in $\overline{\beta}_{ai}^j$.

### 2.4.2 Uncertainty with rules

Although it is usually the case that every effort is made to build a rule-base with certainty as high as possible, experts may not be completely confident as to the rules they are providing. As a consequence, the reasoning system must be capable of dealing with this kind of uncertain knowledge. Assume that a rule $R^j$ is of the form

$$IF\ X_1\ is\ A_1^j\ AND\ X_2\ is\ A_2^j\ AND\ \ldots\ AND\ X_n\ is\ A_n^j \quad THEN\ Y\ is\ B^j\ [\gamma^j] \tag{2.28}$$

where $\gamma^j \in [0,1]$, with the understanding that a completely uncertain rule should be eliminated from the rule-base. Note that the certainty indicator $\gamma^j$ is only attached to the *action* part of the rule. In other words, equation (2.28) states that the only doubt about the rule $R^j$ pertains to the *action* to be taken by $Y$. This kind of rule can be dealt with by modifying "$Y$ is $B^j$ [$\gamma^j$]" into "Y is $\bar{B}^j$", with a certainty of 1 such that

$$\bar{B}^j(v) = Max\ [B(v),\ 1 - \gamma^j] \tag{2.29}$$

When $\gamma^j = 1$, we get $\bar{B}^j = B^j$. Note that as long as $\gamma^j < 1$, the support of $B$ is extended to the whole of $V$, and an element $v \in V$ with grade 0 of membership belonging to $B^j$ now has a grade $1 - \gamma^j$ of membership belonging to $\bar{B}^j$. This means that no value can be totally excluded from $V$. Thus, with this approach the modified rule has the same form as that used in the reasoning model obtained previously and hence the model can be applied without any difficulty.

When both data and rules are considered to be uncertain, the reasoning algorithm (2.23) now can be written as

$$\pi_Y(v) = \underset{k=1}{\overset{K}{\Theta}} \left\{ \left[ \underset{i=0}{\overset{n}{\Xi}} \bar{\beta}_i^k \right] \Phi\ \bar{B}^k(v) \right\} \tag{2.30}$$

We see that the basic structure of the reasoning model is not changed in spite of the fact that uncertainty is present in both data and rules.

## 2.5 Summary

The algorithmic implementation of approximate reasoning has been investigated in this chapter. The resultant models (equations (2.23) and (2.30)) have the following advantageous features:

1. A unified approach. Both fuzzy and random uncertainties are dealt with in a unified way by employing possibility theory.
2. Conceptually clear. Various linguistic connectives are explicitly presented in the models and each of them has a definite interpretation in an algorithmic sense.
3. Very flexible. The change of mathematical definitions of connectives can be accomplished directly in the models without changing the model structure. Therefore it is very convenient to compare these definitions in a unified way.
4. Computation is relatively simple. The calculation procedures are clearly indicated in the models and consequently it can be used for the purpose of on-line reasoning.

CHAPTER 3

# Multivariable blood pressure control: an application of approximate reasoning

---

*A decentralized fuzzy controller structure for dealing with multivariable control of human blood pressure is described and extensive simulations were carried out.*

---

## 3.1 Introduction

Rule-based fuzzy control has been considered as an efficient method to control complicated and ill-defined processes. This results primarily from the fact that no explicit mathematical models governing the controlled process are needed to design the controller. Despite the fact that there are many successful examples found in industrial process control, only a few applications of rule-based control (not necessarily fuzzy) have been reported in biomedical and clinical situations. These situations are very complex, nonlinear and time-varying, and should be suitable candidates for fuzzy control. Vishnoi and Gingrich (1987) constructed a fuzzy logic controller for the delivery of gaseous anaesthesia based on vital signs. Isaka *et al.* (1988, 1989) applied a fuzzy control method to regulate mean arterial blood pressure in a noisy environment. Den Brok and Blom (1987) built a rule-based supervisory expert system to control a patient's blood pressure, in which the expert system was used to tune the parameters of a PID controller based on the analysis of step responses. A hybrid controller for drug delivery was proposed by Neat *et al.* (1989), and consists of a fuzzy controller, a multiple-model controller and a model reference controller. The expert system is used to adjust the structure of the whole controller in accordance with the dynamic plant. Ying *et al.* (1988) and Yamashita *et al.* (1988) have applied fuzzy control method to arterial pressure control by single drug infusion. Linkens *et al.* (1986, 1988) describe an application of fuzzy control to muscle relaxation during surgery.

All the work reviewed above involved the automated infusion of a single drug, which therefore represents a single-input single-output problem. Not surprisingly, these single drug delivery systems have been investigated for some years although the control strategies used have primarily been traditional control algorithms. In contrast, multivariable drug delivery systems have received little attention, and have been studied only recently by some researchers for blood pressure control (Serna *et al.* 1983; McInnis and Deng 1985; Voss *et al.* 1987; and Linkens and Mansour 1989). The control algorithms adopted have been either traditional adaptive schemes or self-tuning control methods.

This chapter deals with the problem of multivariable control of blood pressure which is the same system as used in the work reported by Mansour and Linkens (1990), but here a rule-based fuzzy control method is employed. A control architecture has been developed which consists mainly of a rule-based fuzzy controller plus a simple pre-compensator. The reasoning algorithms used by the fuzzy controller are based on the unified approximate reasoning model derived in Chapter 2. The proposed control method and the reasoning models are investigated via a number of simulations on the models given in Chapter 1. The simulation is aimed at not only demonstrating the feasibility of the proposed method when applied to a relatively complicated situation, but also at evaluating a number of reasoning schemes. Eight reasoning algorithms are chosen for comparison using the following two approaches. The first is to compare the control performance measured by integral of error criteria when the parameters of the controlled model are fixed. The second is to investigate the robustness of the algorithms with respect to variations in each process model parameter. The parameters include pure time delay, time constants, gains and compensation factors.

This chapter is organized as follows. The next section presents an architecture for multivariable fuzzy control. Following this, the simulation results with some analysis and discussion are given. The chapter ends with conclusions.

## 3.2 Architecture of multivariable fuzzy control

In order to handle the difficulty arising from a multivariable process, an architecture for a multivariable fuzzy controller is suggested as shown in Figure 3.1. The system comprises two independent fuzzy controllers $FC_1$ and $FC_1$, a compensator unit, and the process. The main idea is to adopt a decentralized control strategy which treats the multivariable process as two separated single-input single-output processes provided that the input–output pairs are properly selected according to some criteria. The interacting effects are partly removed by introducing a simple compensation procedure which consists of two factors calculated from the approximate steady-state gains of the process. It is also expected that the interaction effect on one loop produced by the other loop, if viewed as a perturbation, can be compensated to some extent by the robustness property possessed by the fuzzy controller. The details concerning the design and implementation are presented below.

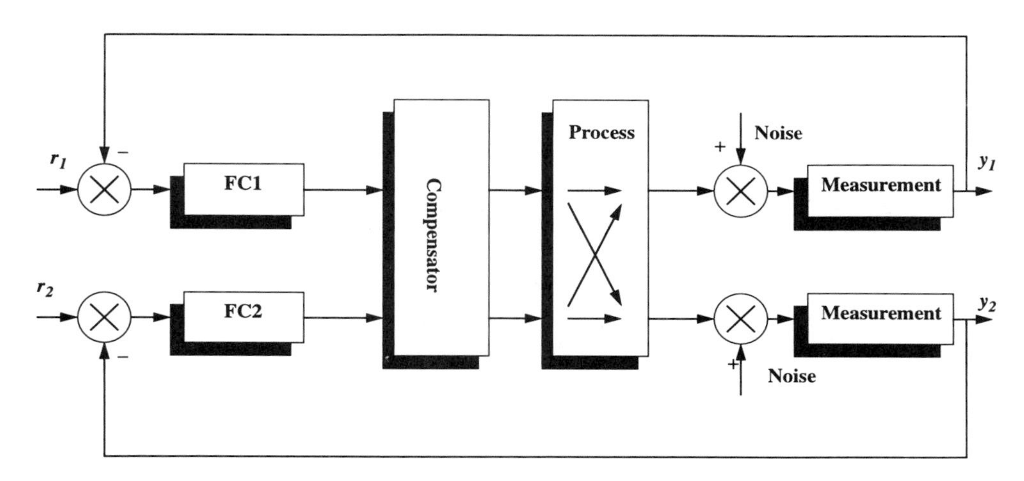


**Figure 3.1** System structure of multivariable fuzzy control

### 3.2.1 Fuzzy control algorithms

The two fuzzy controllers shown in Figure 3.1 have structures depicted in Figure 3.2. It consists of a fuzzification unit, a rule-base, a reasoning algorithm and a defuzzification unit. The fuzzification procedure converts the measured numerical values into fuzzy subsets. The reasoning algorithm infers the corresponding output which is the fuzzy subset, which is subsequently converted into a real numerical control signal by the defuzzification procedure.

#### 3.2.1.1 Constructing the rule-base

Referring to Figure 3.2, suppose that $E$, $C$ and $U$ are variables taking their values on the universes of discourse $\overline{E}$, $\overline{C}$ and $\overline{U}$ respectively. Then a typical rule in the rule-base has the form of "*IF E is* $G_1$ AND C is $G_2$ THEN U is", where $G_1$, $G_2$ and $D$ are the fuzzy subsets defined over the universes of discourse $\overline{E}$, $\overline{C}$ and $\overline{U}$ respectively. They represent some fuzzy concepts such as *positive-big, negative-small and zero* etc. and are numerically characterized by the corresponding membership function $G_1(e)$, $G_2(c)$ and $D(u)$, that is, $G_1(e)$: $\overline{E} \rightarrow [0,1]$, $G_1(c)$: $\overline{C} \rightarrow [0,1]$ and $D(u)$: $\overline{U} \rightarrow [0,1]$. The rule form mentioned above is a mathematical translation of a linguistic statement derived from domain experts. The IF-THEN clause states the relationship between the observed "condition" of the controlled process and the

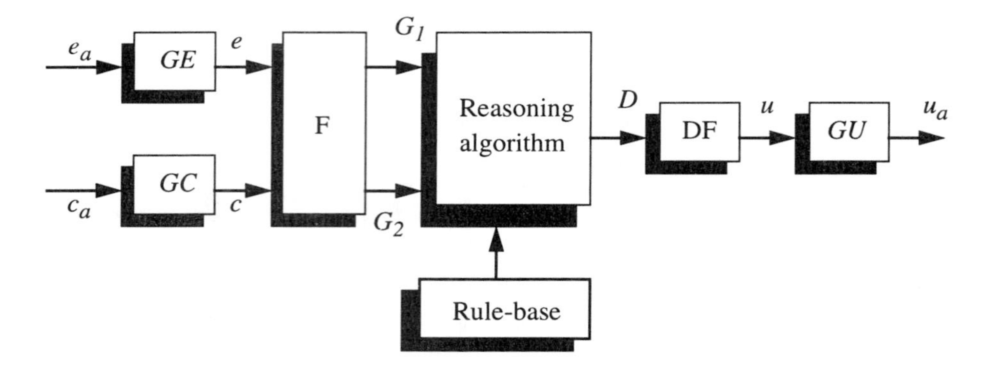


**Figure 3.2** Single loop fuzzy controller

control "action" employed by the expert. An example of this statement might be "IF the error is positive-big AND IF the change-in-error is negative-small, THEN the change-in-control output is positive-medium".

Although it is usually a difficult task to build a good rule-base for a general expert system, the control rules involving a single-input single-output process may be obtained from some common control experience and knowledge. Assume that:

1. The SISO process has a monotonic input–output relationship (not necessarily linear), meaning that the output of the process increases when the input of the process increases or vice versa.
2. The control goal is to maintain a controlled variable at a desired level by manipulating the process input, subject to some disturbance and perturbation.
3. The response of the system will not be ideal due to the inertial property possessed by the process.

Using assumptions similar to those stated above, Yamazaki and Sugeno (1985) obtained a group of control rules that are summarized in Table 3.1, where variables $E$, $C$ and $U$ stand for error, the change-in-error and the change-in-control respectively. NB, NM, NS, ZR, PS, PM and PB represent negative-big, negative-medium, negative-small, zero, positive-small, positive-medium and positive-big respectively. They are fuzzy subsets defined on the corresponding universe of discourse. There are 33 rules in the table with each entry being read as a rule "*IF E is* $G_1$ AND C is $G_2$ THEN U is D".

**Table 3.1** Control rules

| *U* | | Change in error (*C*) | | | | | | |
|---|---|---|---|---|---|---|---|---|
| | | **NB** | **NM** | **NS** | **ZR** | **PS** | **PM** | **PB** |
| **Error (*E*)** | **NB** | | NB | | NM | NM | NS | |
| | **NM** | NB | | NM | | NS | | PS |
| | **NS** | | NM | NS | NS | ZR | PS | PM |
| | **ZR** | NB | | NS | ZR | PS | | PB |
| | **PS** | NM | NS | ZR | PS | PS | PM | |
| | **PM** | NS | | PS | | PM | | PB |
| | **PB** | | PS | PM | PM | | PB | |

### 3.2.1.2 Reasoning algorithm

The main task of the fuzzy controller is to infer the present control output from the present input based on the control rules. This process involves two issues, being the mathematical representation of the rules and the inference implementation. Most existing fuzzy controllers adopt the relation equation method to translate the rules into a matrix and employ Zadeh's inference rule of composition to reach a conclusion. Alternatively, in this work an approximate reasoning model derived in Chapter 2 will be used. It is based on Zadeh's possibility theory rather than on the relation equation method. Suppose that there are $m$ rules in the rule-base, each of which has $n$ input variables and one output variable with the form:

IF $X_1$ is $A_1^j$ AND $X_2$ is $A_2^j$ AND . . . AND $X_n$ is $A_n^j$ THEN Y is $B^j$
Furthermore, assume that the present input is

$X_1$ is $C_1$ AND $C_2$ is $A_2$ AND . . . AND $X_n$ is $C_n$

Then the present output "$Y$ is $D$", represented by its membership function $D(u)$, can be obtained through the following algorithm:

$$D(u) = \overset{m}{\underset{j=1}{\Theta}} \left\{ \left[ \overset{n}{\underset{i=1}{\Xi}} \beta_i^j \right] \Phi \, B^j(u) \right\} \tag{3.1}$$

where "Θ", "Ξ" and "Φ" stand for linguistic connectives ALSO, AND and IF-THEN respectively, and $\beta_i^j$ denotes the possibility measure between $C_i(u_i)$ and $A_i^j(u_i)$ which is defined by

$$\beta_i^j = Poss(C_i|A_i^j) = \underset{u_i \in U_i}{Sup} \, Min \left[ C_i(u_i), A_i^j(u_i) \right] \tag{3.2}$$

Various reasoning algorithms can be obtained by selecting different mathematical definitions for linguistic connectives ALSO, AND and IF-THEN. There exist a number of possibilities but only the four which are used in this study are presented below:

1.

$$D(u) = \bigcup_{j=1}^{m} \left[ \left( \overset{n}{\underset{i=1}{\Lambda}} \beta_i^j \right) \Lambda \, B^j(u) \right] \tag{3.3a}$$

2.

$$D(u) = \bigcup_{j=1}^{m} \left[ \left( \overset{n}{\underset{i=1}{*}} \beta_i^j \right) * B^j(u) \right] \tag{3.3b}$$

3.

$$D(u) = \bigcap_{j=1}^{m} \left\{ 1 \, \Lambda \left[ 1 - \left( \overset{n}{\underset{i=1}{\Lambda}} \beta_i^j \right) + B^j(u) \right] \right\} \tag{3.3c}$$

4.

$$D(u) = \bigcap_{j=1}^{m} \begin{cases} 1 & \text{if } \; [\overset{n}{\underset{i=1}{\Lambda}} \beta_i^j] \le B^j(u) \\ B^j(u) & \text{if } \; [\overset{n}{\underset{i=1}{\Lambda}} \beta_i^j] > B^j(u) \end{cases} \tag{3.3d}$$

where $\bigcup$ and $\bigcap$ correspond to maximum and minimum operators respectively, whereas Λ and * denote minimum and algebraic product operators.

### 3.2.1.3 Computation procedures

The computational process from a measurement $e_a$ and a change-in-error $c_a$ to the controller output $u_a$ can be described as follows.

First, measured numerical values $e_a$ and $c_a$ at sample instant $nT$ are mapped into

$e \in \bar{E}$ and $c \in \bar{C}$ respectively, that is,

$$\begin{cases} e = GE \cdot e_a(nT) \\ c = GC \cdot [e_a(nT) - e_a(nT-T)] \end{cases} \tag{3.4}$$

where $T$ is the sample period, $e_a\ (nT)$ = set-point $r(nT)$ - process output $y(nT)$,and $GE$ and $GC$ are scaling factors which are selected so that all possible measured values of $e_a$ and $c_a$ will fall into the corresponding universes $\bar{E}$ and $\bar{C}$.

Second, the fuzzification unit fuzzifies $e$ and $c$ into fuzzy subsets $G_1$ and $G_2$ that will be singletons characterized by

$$G_1(e) = \begin{cases} 1 & \text{if } e = e(nT) \\ 0 & otherwise \end{cases} \tag{3.5a}$$

$$G_2(c) = \begin{cases} 1 & \text{if } c = c(nT) \\ 0 & otherwise \end{cases} \tag{3.5b}$$

where $e \in \bar{E}$ and $c \in \bar{C}$.

At this point the input data with respect to the reasoning model are of the standard form "$E$ is $G_1$ and $C$ is $G_2$". Based on these data, the controller output "$U$ is $D$" can be inferred by means of one of the algorithms presented in equation (3.3).

The value $D$ obtained from the reasoning stage is a fuzzy subset defined on $\bar{U}$. To implement the control action, a real numerical value is required. This will be done by utilizing a defuzzification stage. In general, there are two popular algorithms called the Mean of Maxima and the Centre of Gravity given by equations (3.6) and (3.7) respectively.

1. The Mean of Maxima (MOM)

$$u = \frac{1}{L} \sum_{l=1}^{L} u_l \tag{3.6}$$

where $u_l$ satisfies $D(u_l) = \underset{u \in \bar{U}}{Max}\, D(u)$.

2. The Centre of Gravity (COG)

$$u = \frac{\sum_{k=1}^{K} u_k \cdot D(u_k)}{\sum_{k=1}^{K} D(u_k)} \qquad u_k \in \bar{U} \quad \text{for } all\ k \tag{3.7}$$

where $K$ is the element number in $\bar{U}$, i.e., Card($\bar{U}$) = $K$.

The final step in the calculation is to convert the integer value $u \in \bar{U}$ computed from one of the above equations into an appropriate control signal $u_a$ via multiplying $u$ by a scaling factor $GU$.

Up to now, the whole computational procedures within a sample period $T$ are carried out if the single-input single-output process is involved, with the understanding that $u_a$ is the change in control. The procedures, however, are not completed when a multivariable process is considered. This will be discussed in the next section.

### 3.2.2 Compensation procedure

As mentioned previously, the interactive effects between the variables should be taken into account. A simple compensator is connected to the outputs of two fuzzy controllers to cope with this problem. It should be emphasized that this scheme does not aim to eliminate the interaction completely. Rather, it is intended to reduce it to acceptable levels using a simple structure.

It is impossible to design a controller, no matter what principle it is based on, without knowing something about the controlled process. The process information needed, however, may be more/less, explicit/implicit, qualitative/quantitative, or exact/vague, depending on the design strategies used. For instance, no explicit and exact process model is needed for designing a rule-based fuzzy controller, because all of the information about the process is fused and embodied into the control rule-base in a qualitative manner. We want to use as little information about the process as possible. It is argued by Grosdidier and Morari (1985) that steady-state information can be obtained for open-loop stable systems and is often the only information available. The steady-state gains in a multivariable system may be represented by a matrix denoted as $G$. In some cases, $G$ can be obtained readily by undertaking experiments on the system considered, or by questioning skilful domain experts.

If, on the other hand, the process is characterized by a set of differential equations (not necessarily linear), $G$ may be derived using the following procedures. Suppose the system is described by

$$\frac{d\mathbf{x}}{dt} = \mathbf{f}(\mathbf{x},\mathbf{u}) \tag{3.8a}$$

$$\mathbf{y} = C\mathbf{x} \tag{3.8b}$$

where $\mathbf{f}$ is the function vector, $\mathbf{x} \in R^n$ is the state vector, $\mathbf{u} \in R^m$ is the control vector, $\mathbf{y} \in R^m$ is the output vector and C is a $m \times n$ constant matrix. An incremental model near the nominal states may be obtained by means of the first-order approximation of the Taylor series, which is given by

$$\frac{d\Delta\mathbf{x}}{dt} = A\,\Delta\mathbf{x} + B\,\Delta\mathbf{u} \tag{3.9a}$$

$$\Delta\mathbf{y} = C\,\Delta\mathbf{x} \tag{3.9b}$$

where

$$A = \frac{\partial \mathbf{f}}{\partial \mathbf{x}^T} = \begin{bmatrix} \frac{\partial f_1}{\partial x_1} & \frac{\partial f_1}{\partial x_2} & \cdots & \frac{\partial f_1}{\partial x_n} \\ \frac{\partial f_2}{\partial x_1} & \frac{\partial f_2}{\partial x_2} & \cdots & \frac{\partial f_2}{\partial x_n} \\ \cdot & \cdot & \cdots & \cdot \\ \cdot & \cdot & \cdots & \cdot \\ \cdot & \cdot & \cdots & \cdot \\ \frac{\partial f_n}{\partial x_1} & \frac{\partial f_n}{\partial x_2} & \cdots & \frac{\partial f_n}{\partial x_n} \end{bmatrix}$$

and

$$B = \frac{\partial \mathbf{f}}{\partial \mathbf{u}^T} = \begin{bmatrix} \frac{\partial f_1}{\partial u_1} & \frac{\partial f_1}{\partial u_2} & \cdots & \frac{\partial f_1}{\partial u_m} \\ \frac{\partial f_2}{\partial u_1} & \frac{\partial f_2}{\partial u_2} & \cdots & \frac{\partial f_2}{\partial u_m} \\ \cdot & \cdot & \cdots & \cdot \\ \cdot & \cdot & \cdots & \cdot \\ \cdot & \cdot & \cdots & \cdot \\ \frac{\partial f_n}{\partial u_1} & \frac{\partial f_n}{\partial u_2} & \cdots & \frac{\partial f_n}{\partial u_m} \end{bmatrix}$$

are constant matrices.
Then the steady-state equation can be written as

$$A\,\Delta\mathbf{x} + B\,\Delta\mathbf{u} = 0$$

or

$$\Delta\mathbf{x} = -\,A^{-1}B\,\Delta\mathbf{u}$$

Thus the steady state output equation is given by

$$\Delta\mathbf{y} = -\,CA^{-1}B\,\Delta\mathbf{u} \tag{3.10}$$

The above equation indicates an incremental relationship between the small inputs $\Delta\mathbf{u}$ and the corresponding small outputs $\Delta\mathbf{y}$ about the nominal states under steady-state conditions. It is clear that the steady-state matrix $G_{m\times m} = -\,C\,A^{-1}\,B$. A special case occurs when the system is governed by a linear state vector equation

$$\frac{d\mathbf{x}}{dt} = A_l\mathbf{x} + B_l\mathbf{u} \qquad \mathbf{y}_l = C_l\mathbf{x}_l$$

The gain matrix will then be $G = -C_l \cdot A_l^{-1} \cdot B_l$. Equivalently, in the linear case, the matrix $G$ is simply equal to $H(0)$ if the transfer function matrix $H(s)$ is given.

There are many useful closed-loop properties which can be extracted from the open-loop steady-state gain matrix (Grosdidier and Morari 1985). For instance, the interaction effects can be measured by the Relative Gain Array (RGA) which is defined from the gain matrix. The relative gain $\gamma_{ij}$ between an input $u_j$ and an output $y_i$ is the ratio of two steady-state gains: gain $g_{ij}$ between $u_j$ and $y_i$ when all loops are open and the gain $\bar{g}_{ij}$ when all outputs except the $i$th output are tightly controlled. The RGA can be calculated using only the steady-state gain matrix $G$. More specifically, $\gamma_{ij}$, the $(i,j)$th element of the RGA matrix $\Gamma$, is expressed in terms of $g_{ij}$ as

$$\gamma_{ij} = g_{ij} \cdot \hat{g}_{ji}$$

where $g_{ij}$ is the $(i,j)$th element of G and $\hat{g}_{ji}$ denotes the $(j,i)$th element of the matrix $G^{-1}$, the inverse of $G$. For a $2 \times 2$ system with a steady-state gain matrix $G$

$$G = \begin{bmatrix} g_{11} & g_{12} \\ g_{21} & g_{22} \end{bmatrix}$$

the corresponding RGA matrix $\Gamma$ is given by

$$\Gamma = \begin{bmatrix} \gamma_{11} & \gamma_{12} \\ \gamma_{21} & \gamma_{22} \end{bmatrix} = \begin{bmatrix} \frac{g_{11}g_{22}}{\Delta} & -\frac{g_{12}g_{21}}{\Delta} \\ -\frac{g_{12}g_{21}}{\Delta} & \frac{g_{11}g_{22}}{\Delta} \end{bmatrix} \tag{3.11}$$

where $\Delta = g_{11}g_{22} - g_{12}g_{21}$.

There exist a number of interesting algebraic properties of the RGA (Grosdidier and Morari 1985; Jensen *et al.* 1986). It is worth noting that if the transfer function matrix is diagonal or triangular, then $\Gamma = I$, where $I$ is the identity matrix. In the former case, the system is said to be "decoupled" while in the latter case it is known as "one-way interactive".

The RGA, introduced as a measurement of interaction, provides guidelines for pairing input–output variables and for predicting when interactions are significant. The input–output pairing is of significant importance if a decentralized control structure is required. This is due to the fact that satisfactory control would be unattainable if an input were to be paired with an output over which the input has little or no influence. Therefore it is desirable to pair the input–output in such a way that the input has as strong influence as possible over the output. This goal can be achieved by pairing those variables whose relative gain is positive and as close to unity as possible. Also, if any $\gamma_{ij}$ is much larger than one or less than zero, pairing the corresponding variables will result in a loop that is difficult to control.

Suppose that the system considered is a $2 \times 2$ system and a steady-state gain matrix $G$ is available. Furthermore, assume that the input–output pairs are appropriately selected and two feedback control loops are constructed. Then the configuration of the proposed compensator is that shown in Figure 3.3. The relationship between the inputs and outputs of the compensator is given by

$$\begin{bmatrix} u_{a1}^* \\ u_{a2}^* \end{bmatrix} = \begin{bmatrix} \lambda_{11} & -\lambda_{12} \\ -\lambda_{21} & \lambda_{22} \end{bmatrix} \begin{bmatrix} u_{a1} \\ u_{a2} \end{bmatrix} = \begin{bmatrix} 1 & -\frac{g_{12}}{g_{11}} \\ -\frac{g_{21}}{g_{22}} & 1 \end{bmatrix} \begin{bmatrix} u_{a1} \\ u_{a2} \end{bmatrix} \tag{3.12}$$

where the $2 \times 2$ matrix $\Lambda$ is called the compensation matrix.

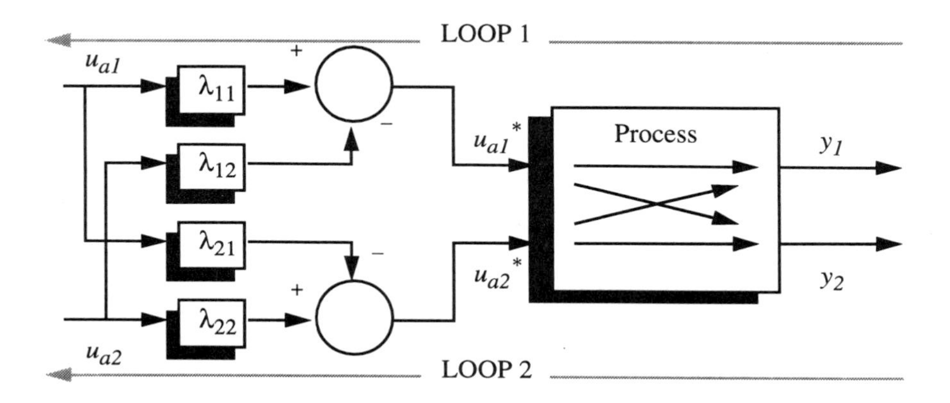


**Figure 3.3** Configuration of compensator

The basic operation of the compensator may be stated as follows. When there is a change in the first controller output $u_{a1}$, it not only produces a change in the input of loop 1, but also presents a change in the input of loop 2 with an amount $-\lambda_{21} \cdot u_{a1}$ so that when the interaction from loop 1 to loop 2 occurs, it will be compensated by the change in the input of loop 2. As a result, the effect caused by the interaction in $y_2$ will be reduced. The same behaviour will take place with respect to the change of the second controller output $u_{a2}$.

The reason for selecting $\lambda_{11} = \lambda_{22} = 1$ is apparent from the above discussion, whereas the determination of $\lambda_{12}$ and $\lambda_{21}$ is based on the following concept. Suppose that there is a change in $y_1$, denoted as $\Delta y_1$, caused by interaction, due to the change of $u_2$, denoted as $\Delta u_2$. $\Delta y_1$ would be approximately equal to $\Delta u_2 \cdot g_{12}$ . This $\Delta y_1$ may be equivalently caused by $g_{11} \cdot \Delta u_1$ assuming that only $u_1$ is applied to the loop 1. Thus to prevent interaction we have

$$g_{11} \cdot \Delta u_1 = -\Delta u_2 \cdot g_{12}$$

or

$$\Delta u_1 = (-\frac{g_{12}}{g_{11}}) \cdot \Delta u_2 = \lambda_{12} \cdot \Delta u_2 \tag{3.13}$$

Therefore, $\lambda_{12}$, in fact, is the ratio of two inputs provided that only loop 1 is concerned. It indicates what percentages of $\Delta u_2$ should be applied to the input of loop 1 in order to compensate for the perturbation caused by $\Delta u_2$. Likewise, $\lambda_{21}$ can be determined by the same consideration.

This result can also be verified by a simple calculation. Let us suppose that the system is in steady-state. Then the following equation holds:

$$\begin{bmatrix} y_1 \\ y_2 \end{bmatrix} = \begin{bmatrix} g_{11} & g_{12} \\ g_{21} & g_{22} \end{bmatrix} \begin{bmatrix} u_{a1}^* \\ u_{a2}^* \end{bmatrix} = \begin{bmatrix} g_{11} & g_{12} \\ g_{21} & g_{22} \end{bmatrix} \begin{bmatrix} 1 & \lambda_{12} \\ \lambda_{21} & 1 \end{bmatrix} \begin{bmatrix} u_{a1} \\ u_{a2} \end{bmatrix}$$

$$= \begin{bmatrix} g_{11}+g_{12}\lambda_{21} & g_{12}+g_{11}\lambda_{12} \\ g_{21}+g_{22}\lambda_{21} & g_{22}+g_{21}\lambda_{12} \end{bmatrix} \begin{bmatrix} u_{a1} \\ u_{a2} \end{bmatrix} \tag{3.14}$$

Let

$$g_{11}\lambda_{12} + g_{12} = 0$$

$$g_{22}\lambda_{21} + g_{21} = 0$$

We obtain

$$\lambda_{12} = -\frac{g_{12}}{g_{11}} \qquad \lambda_{21} = -\frac{g_{21}}{g_{22}}$$

These are identical to the results derived previously from a conceptual consideration.

It is clear that the two components $\lambda_{12}$ and $\lambda_{21}$ are essentially two feedforward control elements, the compensation action taking place before the perturbation produced by the other input enters the loop being considered. It should be stressed that because the parameters of the compensator are derived only from the steady-state gains, which may be inaccurate or time-varying, we never expect the interaction to be removed completely in both transient and steady-states. Instead, what we want is to reduce the effects partly by means of a simple and realizable structure. The rest of the effects are intended to be compensated by the robustness of the fuzzy controller with respect to the perturbation.

Notice that the $u_{a1}$ and $u_{a2}$ are the changes in the control at a specific sample time for each loop. The real control value for each loop is given by

$$\begin{cases} i_{a1}(nT) = i_{a1}(nT-T) + u_{a1} \\ i_{a2}(nT) = i_{a2}(nT-T) + u_{a2} \end{cases} \tag{3.15}$$

## 3.3 Simulation results

There are eight algorithms if we combine the four reasoning methods presented in equation (3.8) with the two defuzzification algorithms given in equations (3.6) and (3.7). All the algorithms will be investigated in this study. For simplicity, each algorithm is denoted by one of the following abbreviations: MMM, MMC, PPM, PPC, MZM, MZC, MGM and MGC, in which the first letter stands for the operation of AND with M being minimum and P algebraic product; the second letter denotes the operation of IF-THEN with M being minimum, P algebraic product, Z Zadeh's implication operator and G the Godel implication operator; the third letter designates the method of defuzzification with M being Mean of Maxima and C Centre of Gravity. For instance, the MGC indicates a combination of equation (3.3d) with equation (3.7).

The variables $E$, $C$ and $U$ take values on the same finite and discrete universes of discourse, that is, $\bar{E} = \bar{C} = \bar{U} = \{-6, -5, -4, -3, -2, -1, 0, 1, 2, 3, 4, 5, 6\}$. As mentioned previously (Table 3.1), there are seven fuzzy subsets arranged from NB to PB. The definitions of the grade of membership for these fuzzy sets on the three universes $\bar{E}$, $\bar{C}$ and $\bar{U}$ are taken to be the same. Figure 3.4 shows one of them. The same definitions will be used in all simulation cases in this chapter.

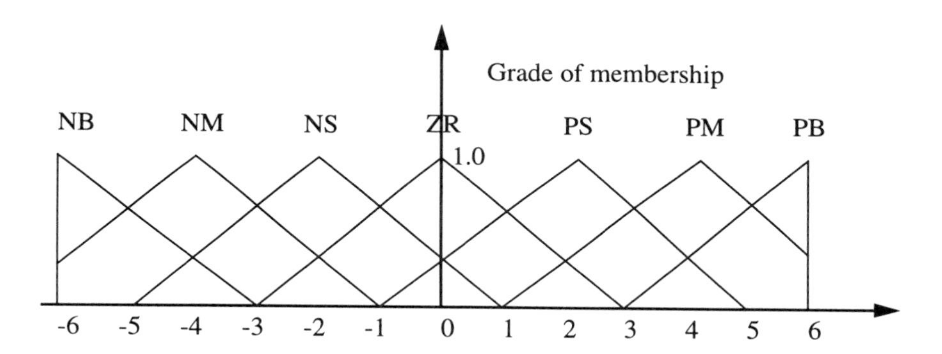


**Figure 3.4** Definition of the membership function

It has been found, from both physiological and mathematical model studies, that the two cases we are dealing with, blood pressure control with and without the drug infusions, approximately satisfy the assumptions made in deriving the rules as summarized in Table 3.1. Therefore, this rule-base will be employed throughout the studies in this chapter.

The objectives of the simulation are to demonstrate the feasibility of the proposed controller when applied to multivariable biomedical situations and to compare various reasoning schemes from a practical viewpoint. The comparative studies are carried out in two senses. One is a comparison of performance, meaning that the controlled responses under various reasoning algorithms were investigated assuming that both controller parameters and process parameters are fixed. The performances were evaluated by two frequently used error integral criteria, i.e. integral of square of the error ($ISE$) and integral of time and absolute error product ($ITAE$), which are defined as

$$ISE = \int_0^\infty e_a^2(t)\,dt$$
$$ITAE = \int_0^\infty t\,|e_a(t)|\,dt \tag{3.16}$$

where $e_a(t)$ is the measured error.

The calculation in the studies was implemented by substituting an algebraic sum for the integrals. The other comparison is via performance robustness, where the aim was to study the robustness property of the proposed controller with respect to variations in the parameters of the process.

### 3.3.1 *PAS* and *PVS* control

The Moller linearized model given by equation (A.3) in Appendix I was used to simulate the cardiovascular system. It was implemented by using a fourth-order Runge Kutta routine with an integration interval of 0.001 s. The controlled variables were $PAS$ and $PVS$ and the manipulated variables were heart rate $HR$ and systemic resistance $RA$.

The steady-state gain matrix $G$ was obtained by using equations (A.3) and (3.10) with outputs being $\Delta PAS$ and $\Delta PVS$, that is,

$$\begin{bmatrix} \Delta PAS \\ \Delta PVS \end{bmatrix} = \begin{bmatrix} 26.69 & 77.25 \\ -0.117 & -1.16 \end{bmatrix} \begin{bmatrix} \Delta HR \\ \Delta RA \end{bmatrix}$$

The RGA matrix $\Gamma$ can be calculated from the $G$ matrix as

$$\Gamma = \begin{bmatrix} 1.41 & -0.41 \\ -0.41 & 1.41 \end{bmatrix}$$

Because $\gamma_{12}$ ($\gamma_{21}$) is negative and $\gamma_{11}$ ($\gamma_{22}$) is positive, proper pairing can be made with $HR/PAS$ and $RA/PVS$. Thus two decentralized control loops can be constructed. In addition, the compensation factors are obtained from the $G$ matrix according to equation (3.12).

Taking a sampling period of 0.1s, the output responses obtained under each approximate reasoning scheme are shown in FigureS 3.5 to 3.8. In the figures, the set-points of $\Delta PAS$ and $\Delta PVS$ were changed at time zero from 0 to 5 mmHg and from 0 to –0.15 mmHg respectively. In order to demonstrate the interactive effects in one loop upon the other and to show the controller's compensating ability, the set-point of $\Delta PAS$ was increased at the 30th sample instant whereas the set-point of $\Delta PVS$ was unchanged. Furthermore, the set-point of $\Delta PVS$ was decreased at the 70th sample instant whereas the set-point of $\Delta PAS$ remained at its previous level.

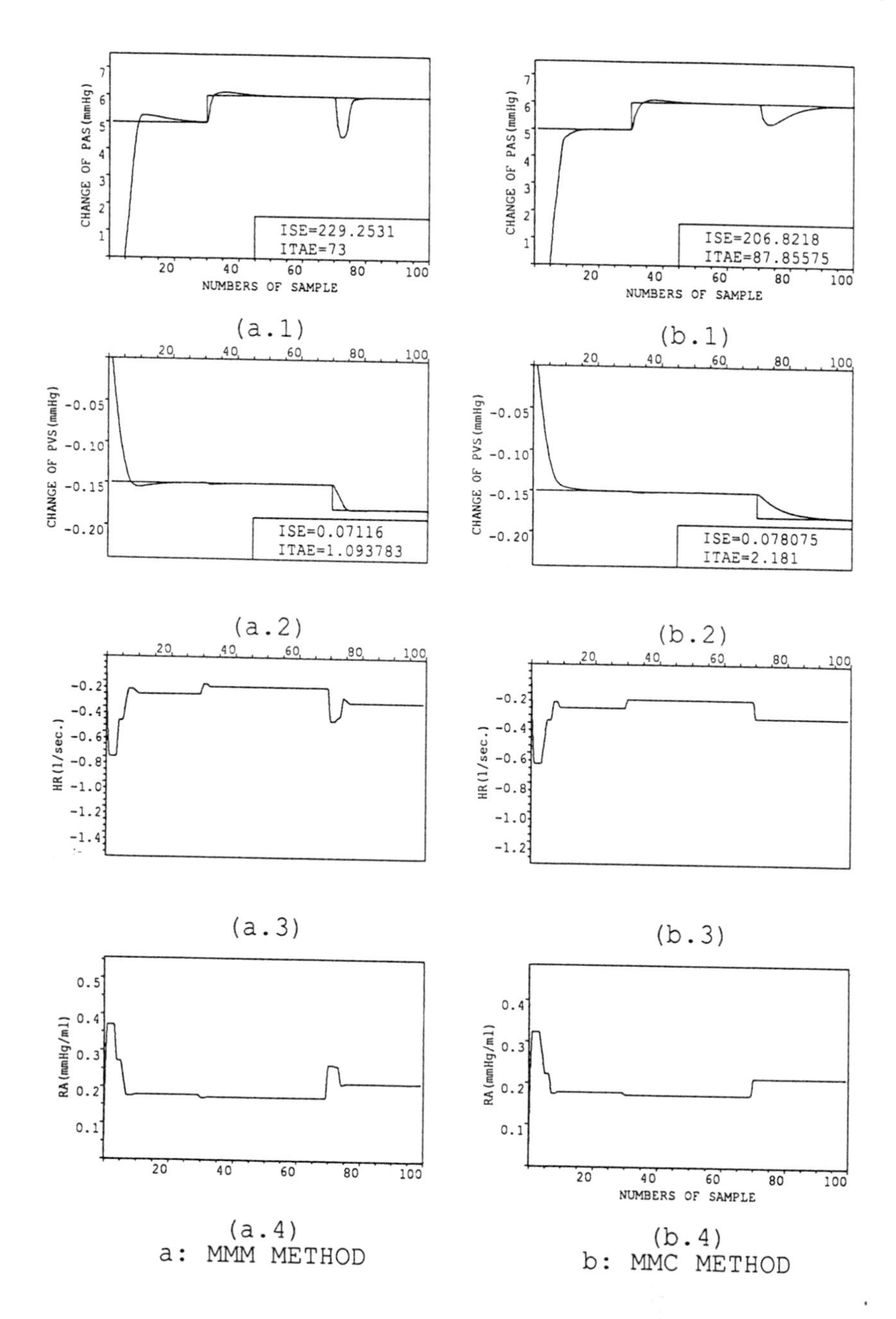


**Figure 3.5** Multivariable fuzzy control of *PAS* and *PVS*: (a) MMM method; (b) MMC method

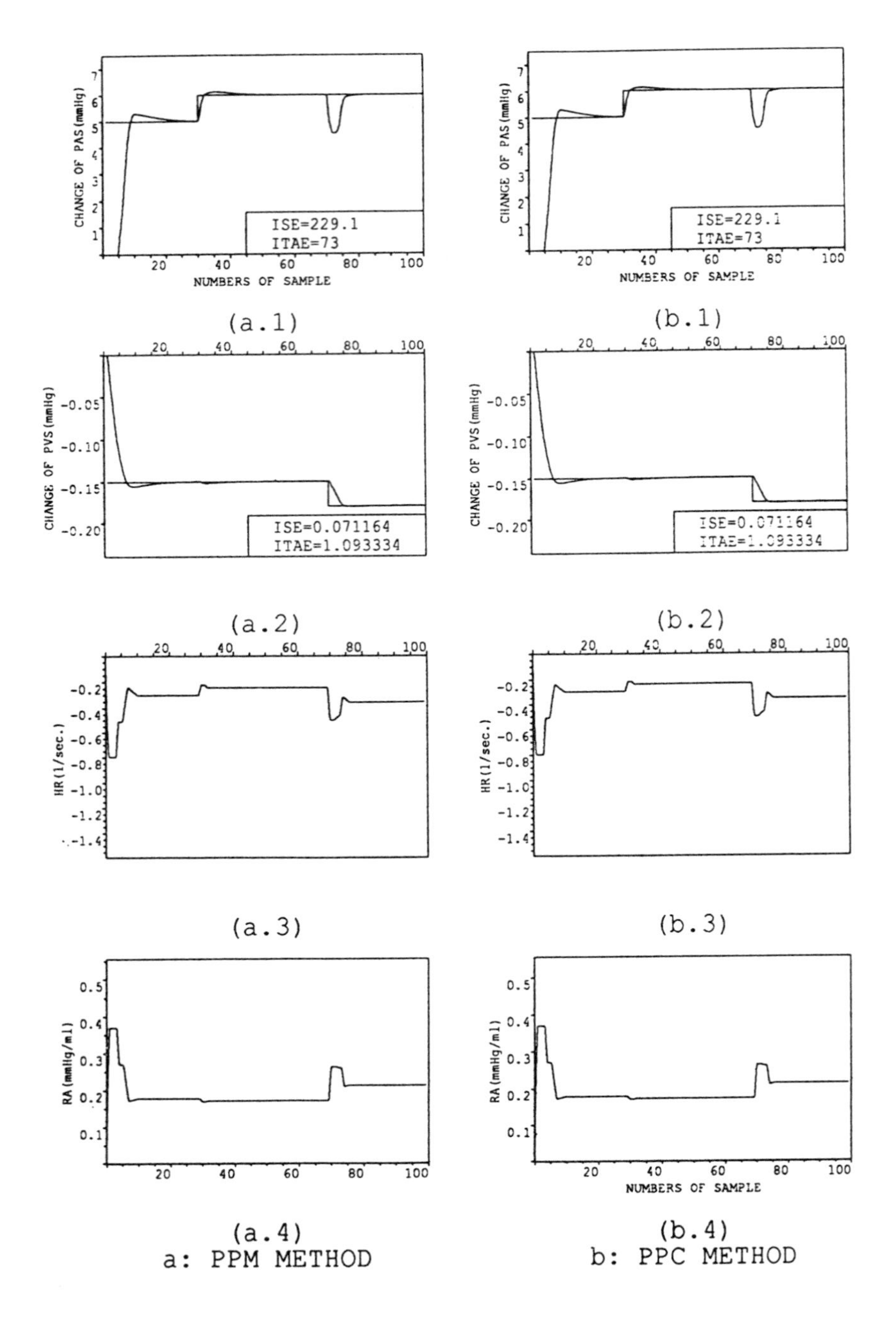


**Figure 3.6** Multivariable fuzzy control of *PAS and PVS*: (a) PPM method; (b) PPC method

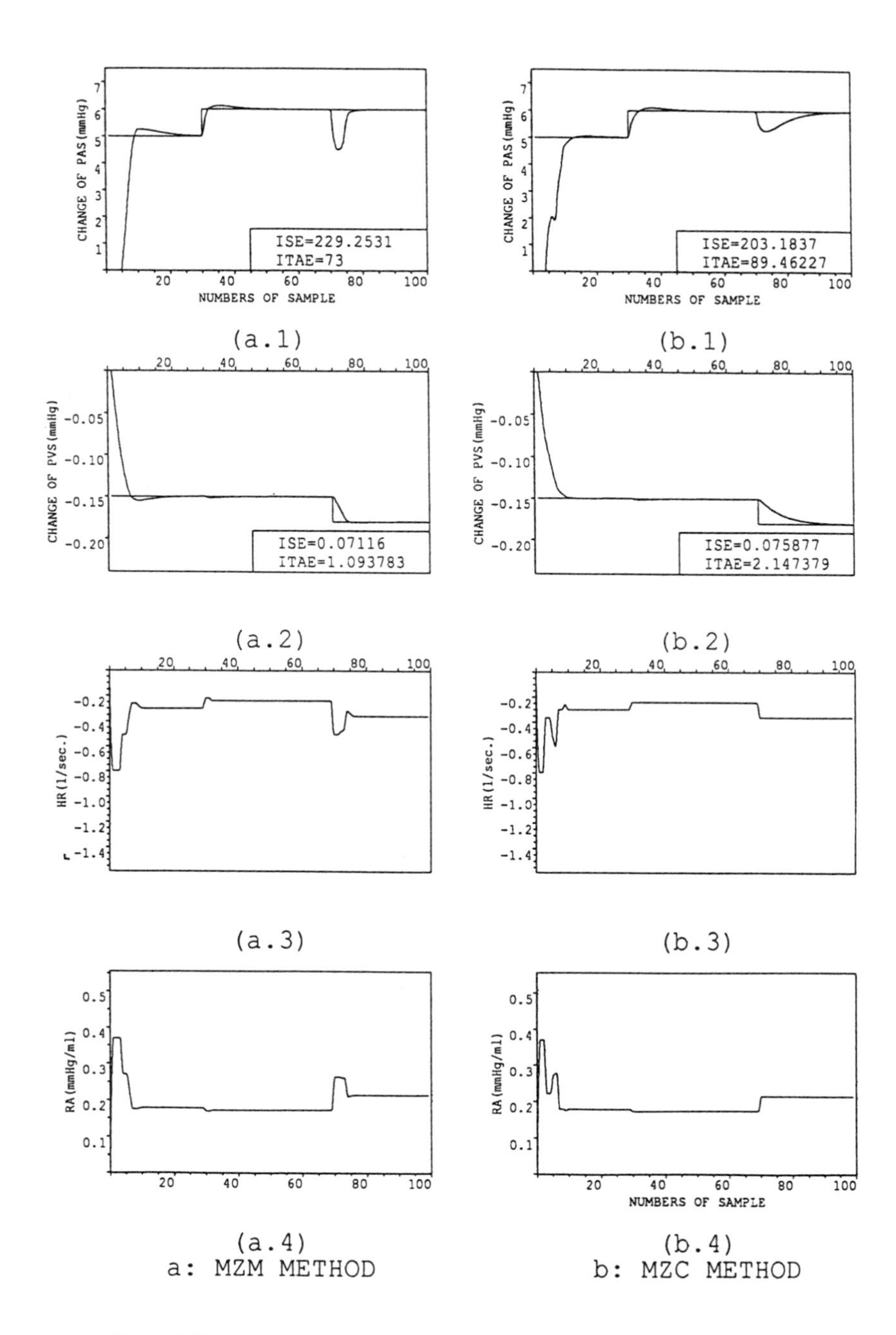


**Figure 3.7** Multivariable fuzzy control of *PAS* and *PVS*: (a) MZM method; (b) MZC method

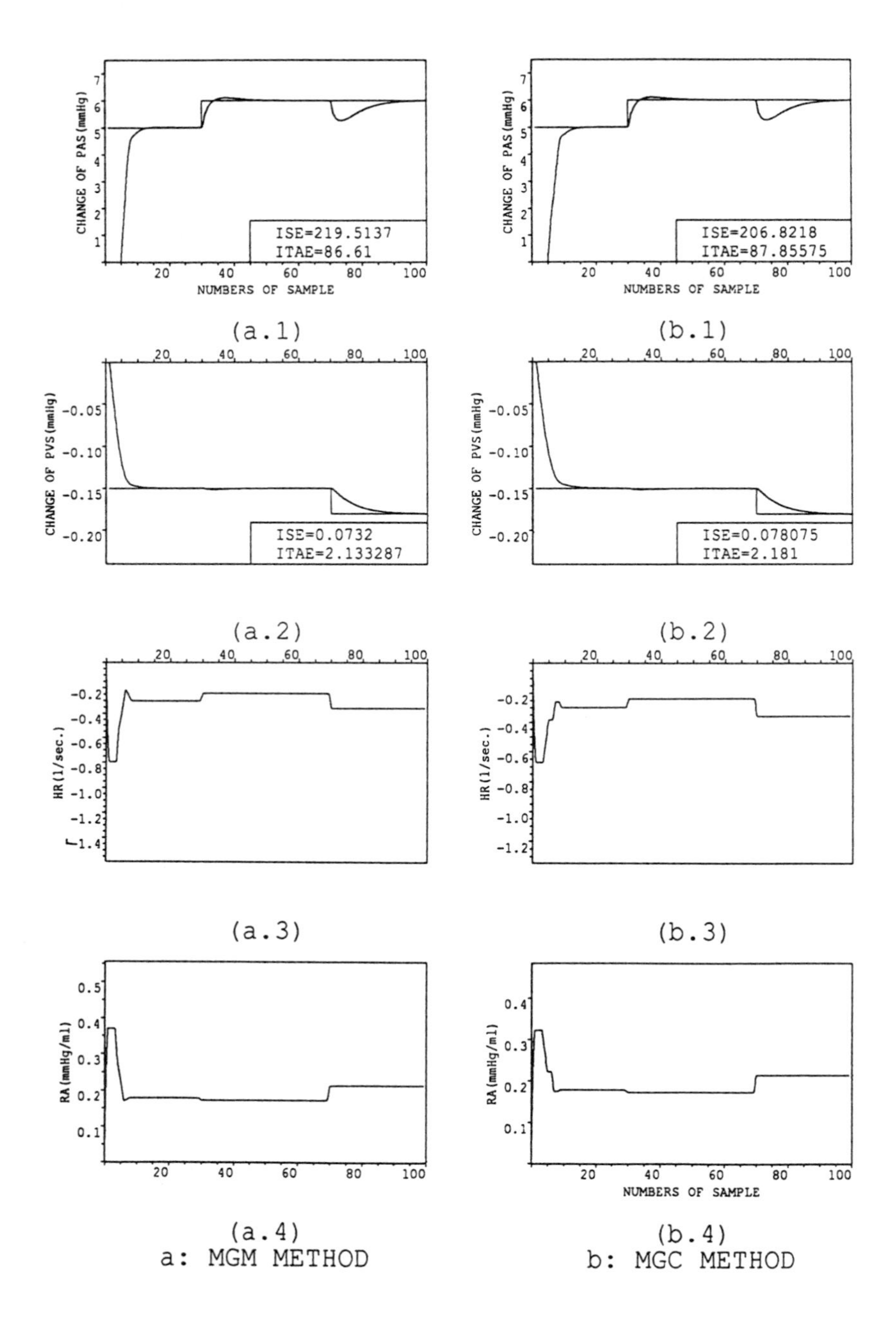


**Figure 3.8** Multivariable fuzzy control of *PAS* and *PVS*: (a) MGM method; (b) MGC method

It can be seen that *PAS* and *PVS* were successfully controlled using the various reasoning approaches. However, there are some differences between the algorithms. While smoother responses with slower tracking speeds were produced by the Godel IF-THEN operator (MGM and MGC), opposite results were obtained by the Minimum and Zadeh implication operators with the use of the Mean of Maxima defuzzification method. As expected, the Centre of Gravity defuzzification strategy usually gave a much smoother transient response than for the Mean of Maxima. An exception is in the case of employing the PPM and PPC algorithms (Figure 3.6), in which algebraic product operators were used for both AND and IF-THEN connectives. It appears that there is no significant difference between the two defuzzification methods in the sense of control performance.

The interactive effects between the two loops are evident in the figures. They appeared at the 30th sampling instant in the *PVS* response and at the 70th sampling instant in the *PAS* response. It can be seen that the proposed controller was able to handle the interaction problem satisfactorily. Note that the two effects were different in amplitude. The change of set-point for *PVS* had a much greater influence on the *PAS* loop than did the change in *PAS*. This is primarily due to the fact that *RA* had a stronger cross-coupling effect on PAS than that produced by *HR* on *PVS*.

### 3.3.2 Simultaneous control of *CO* and *MAP*

As discussed in Appendix II, it is necessary to increase *CO* while *MAP* is lowered. Equation (A.5) is an approximate model, consisting of the *CVS* and the drug dynamics, which relates *CO* and *MAP* to the infusions of DOP and SNP. When $K$, $\tau$ and $T$ are set to the typical values given in Appendix II, the steady-state gain matrix can be calculated as

$$G = \begin{bmatrix} 1.0 & -24.76 \\ 0.6635 & 76.38 \end{bmatrix} \begin{bmatrix} 8.44 & 5.275 \\ -0.09 & -0.15 \end{bmatrix}$$

and the RGA matrix $\Gamma$ is given by

$$\Gamma = \begin{bmatrix} 1.559 & -0.1559 \\ -0.1559 & 1.559 \end{bmatrix}$$

Considering the elements of $\Gamma$, we conclude that the loops should be constructed by pairing *CO*/DOP and *MAP*/SNP. It is worth noting that this conclusion agrees with the conceptual consideration in which DOP is primarily used to increase *CO*, and SNP is mainly aimed at lowering *MAP*.

The same controller parameters were used as in the previous simulations, except for the scaling factors and compensator elements. In the first place, the parameters of the simulated drug dynamic model were fixed at typical values. Again, eight algorithms were investigated. The sampling period was chosen to be 30 s. The results obtained are shown in Figures 3.9 to 3.12. In each figure two groups of results named a and b are displayed which were obtained by using two different algorithms in which the defuzzification method was different whereas the reasoning

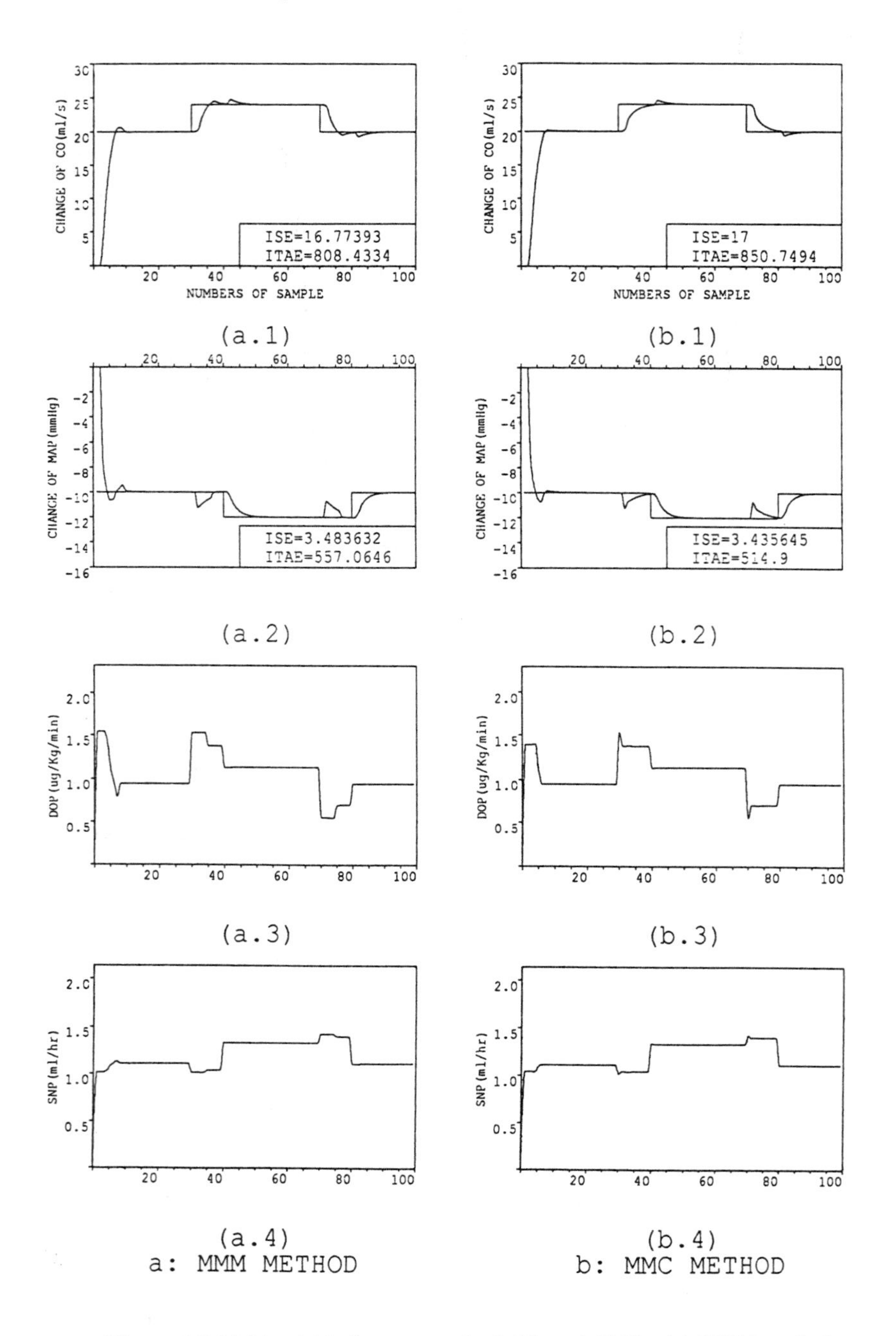


**Figure 3.9** Multivariable fuzzy control of *CO* and *MAP*: *(a) MMM method;* (b) MMC method

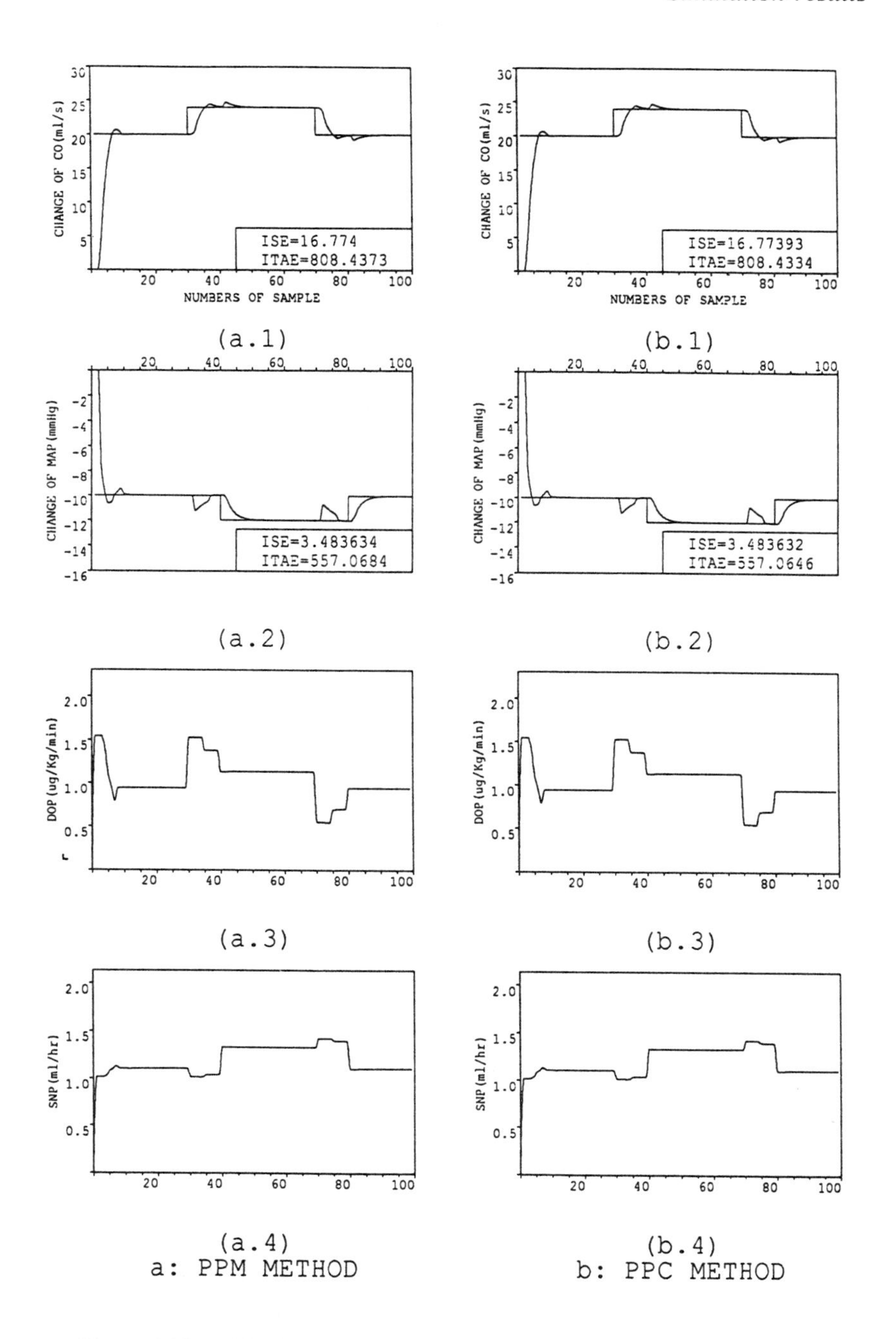


**Figure 3.10** Multivariable fuzzy control of *CO* and *MAP*: (a) PPM method; (b) PPC method

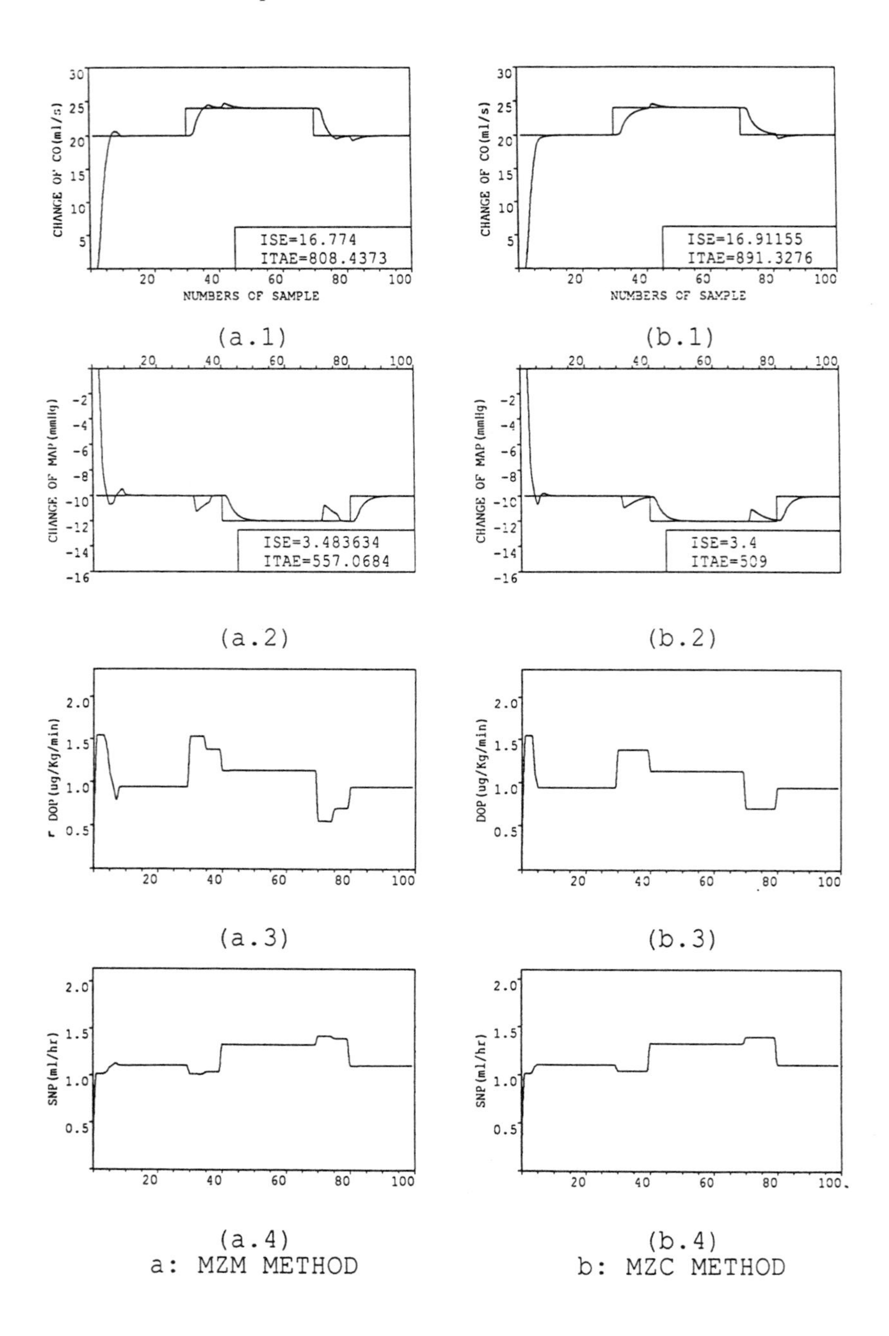


**Figure 3.11** Multivariable fuzzy control of *CO* and *MAP*: (a) MZM method; (b) MZC method

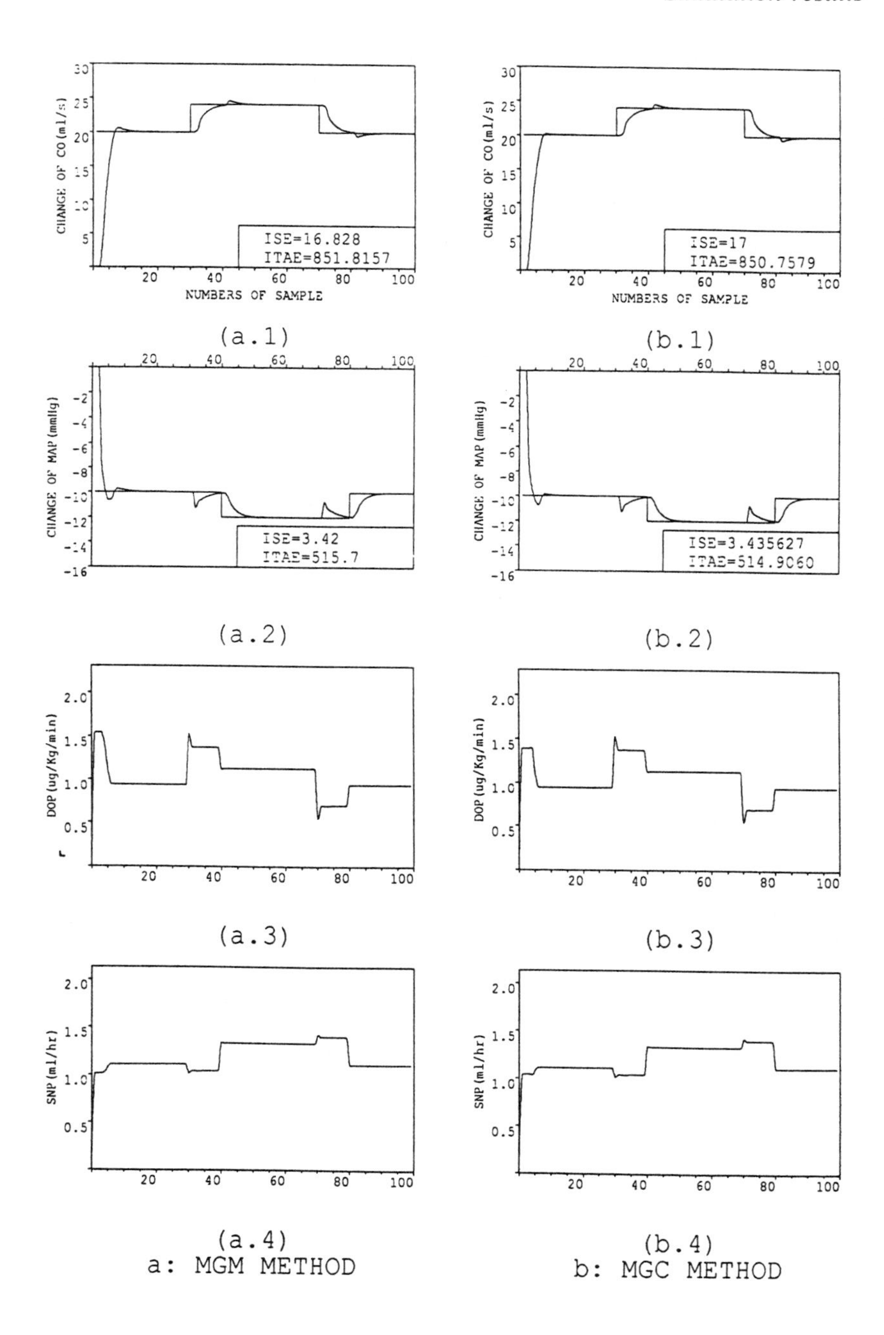


**Figure 3.12** Multivariable fuzzy control of *CO* and *MAP*: (a) MGM method; (b) MGC method

scheme remained the same. The set-points for *CO* and *MAP* were square-like but with different times of change in order to show the interactive effects and to demonstrate the compensating ability possessed by the proposed controller.

It is clear that all the eight algorithms could be used to control the process. Comparing these figures, it can be seen that there are no large differences within the reasoning algorithms although the integral indices are slightly different. However, different transient responses were exhibited if the defuzzification method was different under the same reasoning scheme. This can be seen by comparing the two responses in the same figure. A smoother response was given by means of the Centre of Gravity method.

Next, background noise was added to the simulation model. The simulated noise, which was uniformly distributed in the interval of $-1$ and $+1$, is shown in Figure 3.13. The amplitude of the noise was about 20% of the set-points. Because of space limitation, only one of the results is shown in Figure 3.14. This was obtained by using MMM and MMC algorithms. The result indicates the feasibility of the controller within a noise-contaminated environment. Basically, the output responses could follow the change of the set-point and provide good regulation despite the presence of the noise. The performance of the system would be improved further if the noise-contaminated measurement data were filtered before being used for feedback.

Finally, the robustness of specific reasoning algorithms with respect to changes in the process parameters was investigated. Here the robustness is referred to as performance robustness. Comparative studies on eight algorithms were performed according to the following steps. First, a set of typical process parameters was selected known as the reference parameter set $\Omega_r \in \Omega = \{T_1, T_2, \tau_1, \tau_2, g_{11}, g_{12}, g_{21}, g_{22}\}$ with $T$, $\tau$ and $g$ being the time constants, the transport delays and the gains. Second, eight algorithms were employed in turn with the controller parameters taken to be the same for all algorithms. The results produced were evaluated by a set of indices known as the reference index set $\Psi_r \in \Psi = \{$CISE, PISE, CITAE, PITAE, SISE, SITAE$\}$, where the first letter in each element stands for the different response, that is, C is for *CO*, P for *MAP* and S for the sum of the corresponding two indices (e.g. SISE = CISE + PISE). Note that in this way eight reference index sets were generated. Then one of the process parameters was varied, and the eight algorithms were applied again without changing any controller parameters. Thus eight index sets were obtained which will be compared with the corresponding reference index set.

Let us define the relative variation with respect to a specific process parameter $\omega \in \Omega$ as

$$\tilde{\Delta\omega} = \frac{\Delta\omega}{\omega_r} = \frac{|\omega - \omega_r|}{\omega_r} \tag{3.17}$$

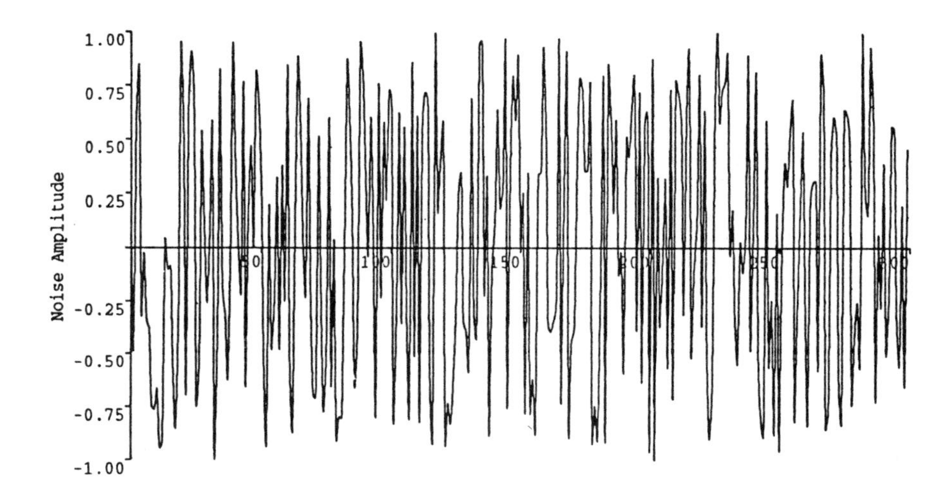


**Figure 3.13** Random background noise

with $\omega_r \in \Omega_r$ and relative variation with respect to a specific control index $\psi \in \Psi_r$ as

$$\tilde{\Delta\psi} = \frac{\Delta\psi}{\psi_r} = \frac{|\psi - \psi_r|}{\psi_r} \tag{3.18}$$

with $\psi_r \in \Psi_r$.

It is evident that the smaller the ratio of $\Delta\tilde{\psi}/\Delta\tilde{\omega}$, the more robust is the corresponding reasoning algorithm with respect to this parameter, based on this performance index.

Figure 3.15 shows the results obtained when the gains of the drug dynamic model represented by equation (1.5), $k_{11}$, $k_{12}$, $k_{21}$ and $k_{22}$, were varied by 10% about their typical values. The horizontal axis represents eight reasoning algorithms whereas the vertical axis corresponds to one of the six relative indices defined by equation (3.18). Actual responses are not presented here because of the huge number of simulation results and space limitation. Generally, the responses were acceptable, especially for transient behaviour. It should be noted, however, that sometimes a non-zero steady-state error appeared which resulted in a relatively big *ITAE*.

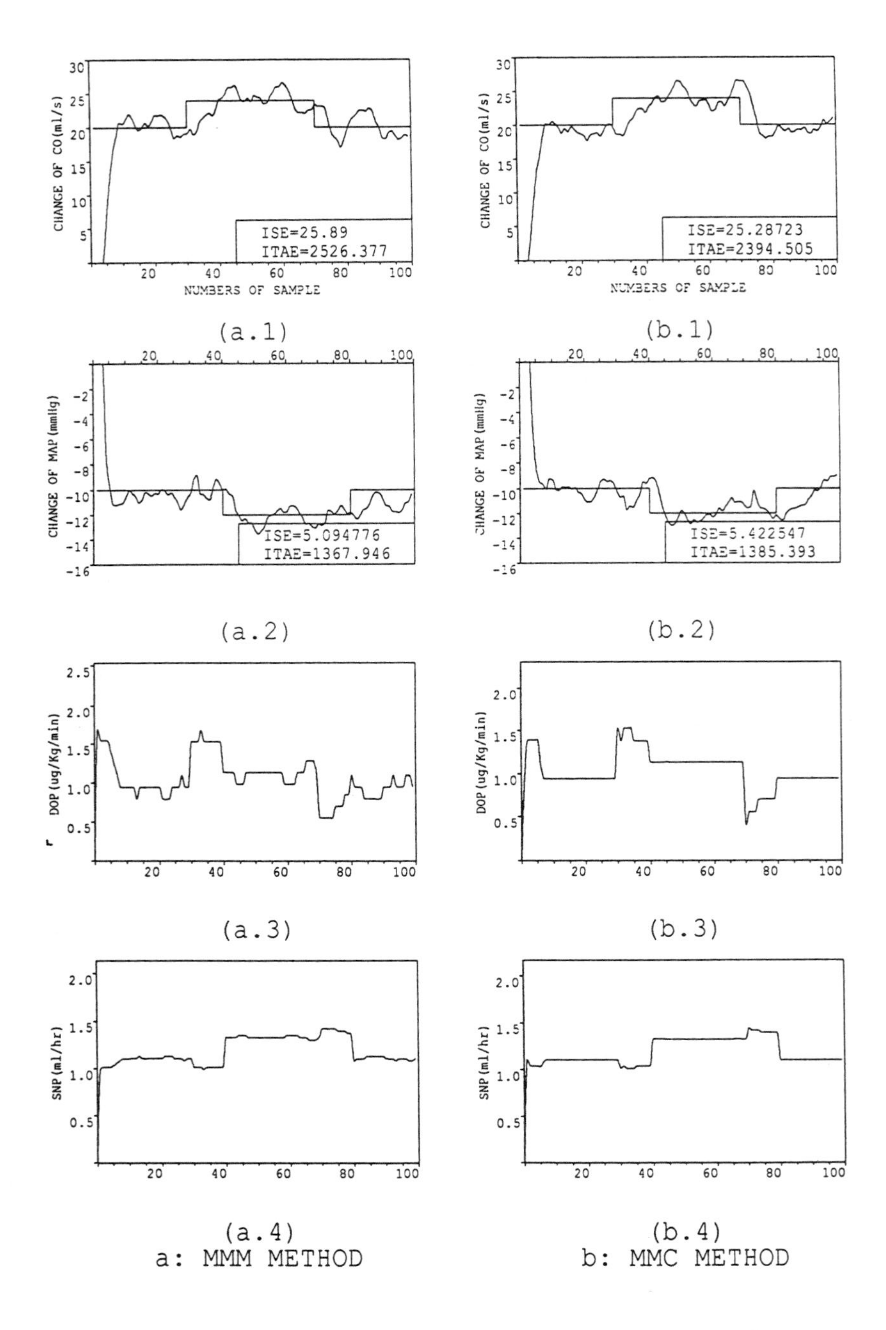


**Figure 3.14** Fuzzy control of *CO* and *MAP* with background noise

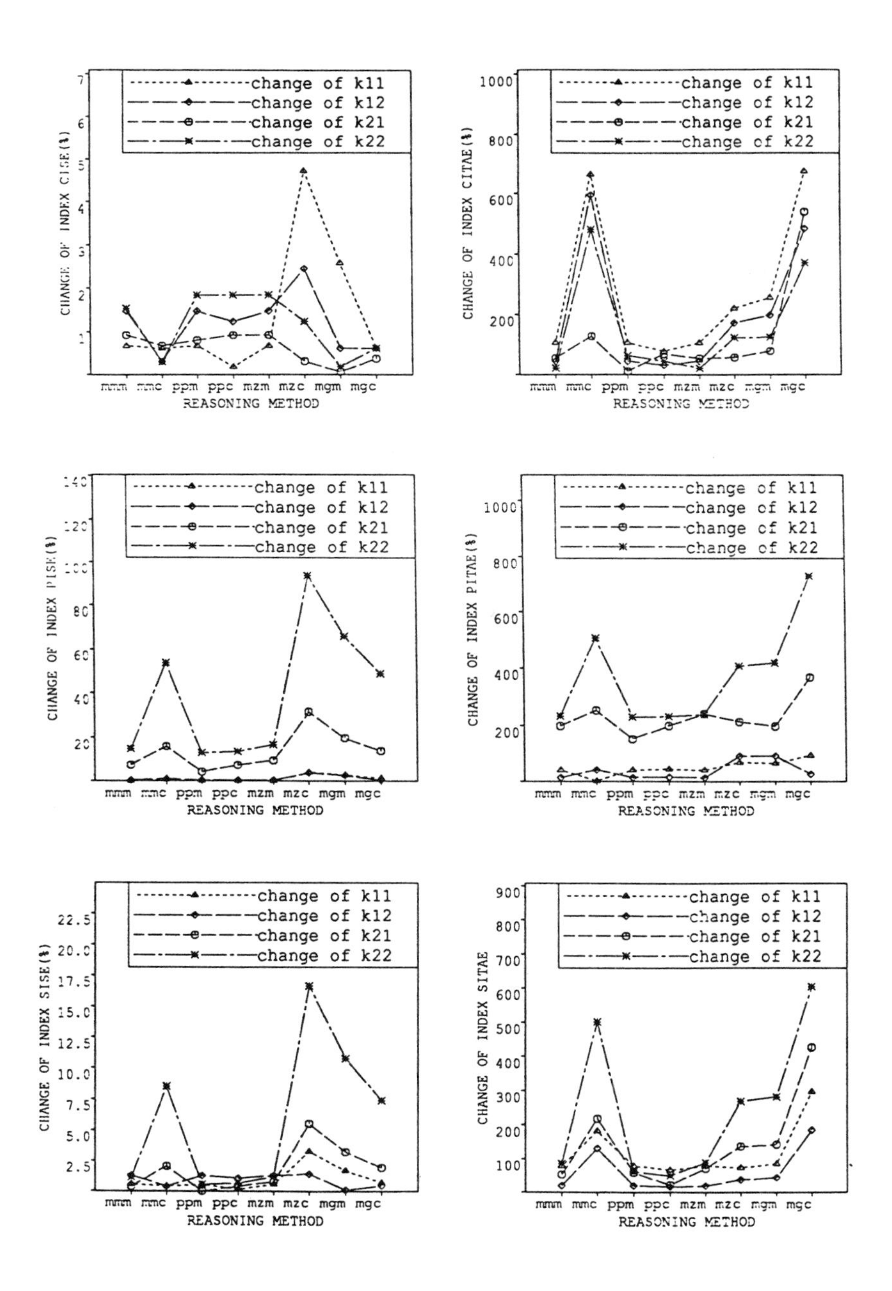


**Figure 3.15** Robustness to change in the gains

It appears that the COG defuzzification method is less robust than the MOM. This can be seen from the figures. The same reasoning algorithm with the COG method usually produced larger index variations than with the MOM method. This may be explained by the fact that the former pays more attention to details of the inferred fuzzy subset than the later.

We can deduce, by careful observation and comparison, that Mamdani's and Larsen's IF-THEN operators (minimum and algebraic product) are usually less sensitive to gain variation than Zadeh's and Godel's IF-THEN operators. It should be noted that an exception is the MMC algorithm which usually gave much worse response and therefore bigger error indices than the others, although this algorithm is commonly used in the literature. It is also worth noting that almost all of the algorithms were less robust with respect to variation in the gains $k_{21}$ and $k_{22}$ than for the gains $k_{11}$ and $k_{12}$. In other words, not all gains have equal influence on the robustness property of the algorithms.

The comparative results when variations in the time constants and the pure time delays were made are shown in Figure 3.16. The time constants $T_1$ and $T_2$ and time delays $\tau_1$ (d1 in Figure 3.16) and $\tau_2$ (d2 in Figure 3.16) were changed by 10% about their typical values. It seems that most conclusions for the case of gain variation with respect to the algorithms are applicable also to this situation. However, we notice that changing the time constants gave rise to a small variation in the indices. In contrast, changing the time delays usually resulted in a larger change in the indices, thus demonstrating large sensitivity to these parameters.

The final simulation results involved the investigation of effects caused by changes in the compensator factors, namely $\lambda_{12}$ and $\lambda_{21}$ in equation (3.14). The purpose was to evaluate how robust the proposed multivariable fuzzy controller is with respect to the compensator factors, as indicated by the error integral indices. We considered four cases, in each of which $\lambda_{12}$ and $\lambda_{21}$ were changed by a different amount. If the nominal values of $\lambda_{12}$ and $\lambda_{21}$ are denoted as $\lambda_{12}^*$ and $\lambda_{21}^*$ and their changed values are $\lambda_{c12}$ and $\lambda_{c21}$, then four cases were considered as follows:

Case 1: $\lambda_{c12} = 1.5*\lambda_{12}^*$ $\lambda_{c21} = 0.5*\lambda_{21}^*$
Case 2: $\lambda_{c12} = 0.5*\lambda_{12}^*$ $\lambda_{c21} = 1.5*\lambda_{21}^*$
Case 3: $\lambda_{c12} = 0.0*\lambda_{12}^*$ $\lambda_{c21} = 0.0*\lambda_{21}^*$
Case 4: $\lambda_{c12} = 0.75*\lambda_{12}^*$ $\lambda_{c21} = 0.75*\lambda_{21}^*$

It should be noted that the variations of $\lambda_{12}$ and $\lambda_{21}$ are due to changes in the steady-state gains. Therefore the cases we are considering are, in fact, intended to show the robustness performance of the compensator factors with respect to the variation of the steady-state gains. This variation may reflect inconsistency between the estimated and actual gains.

The simulation results are shown in Figure 3.17. It is clear that the worst situation occurred when case 3 was tried. In this case there is no compensation action applied to the system, and as expected, the interactive effects within the two loops

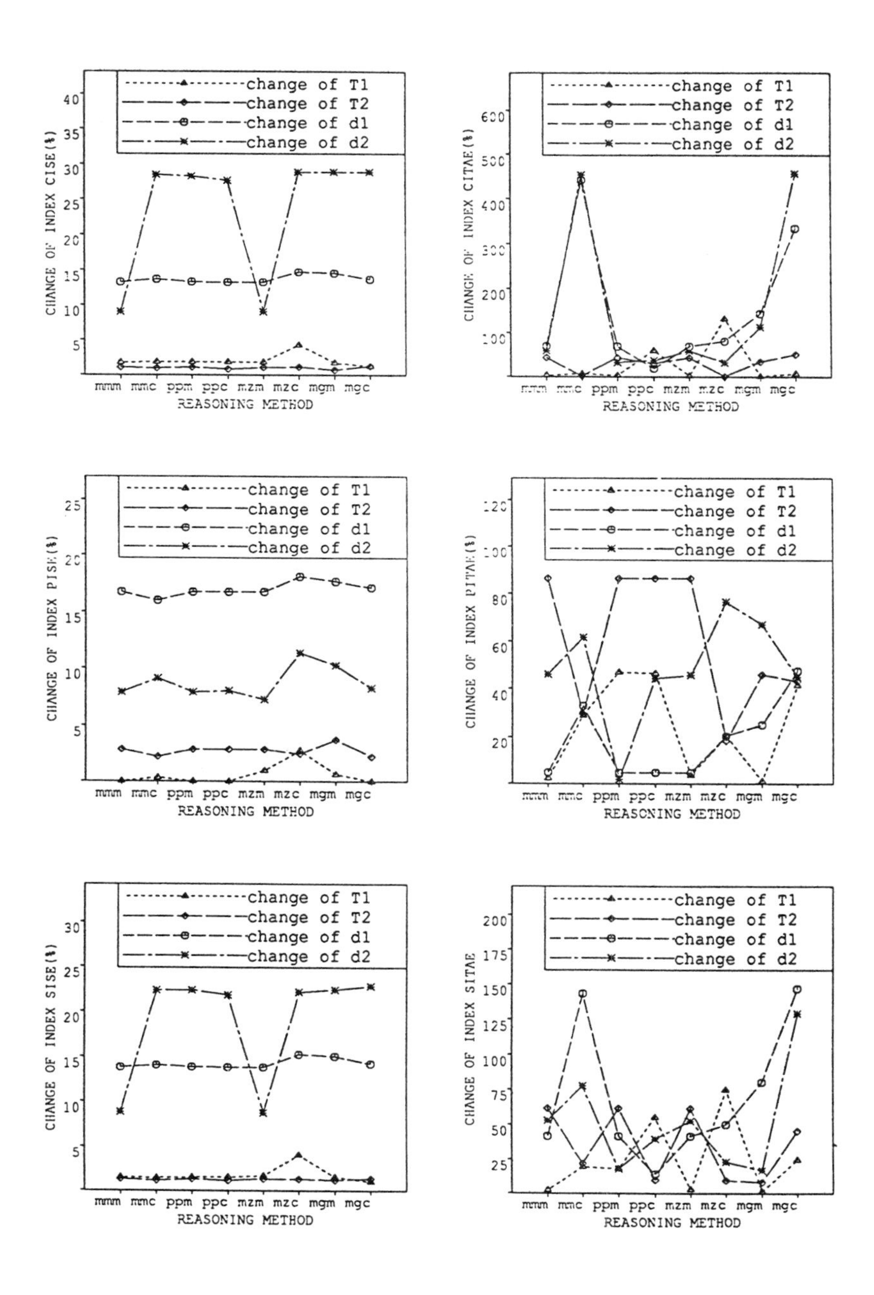


**Figure 3.16** Robustness to change in the time-constants and delays

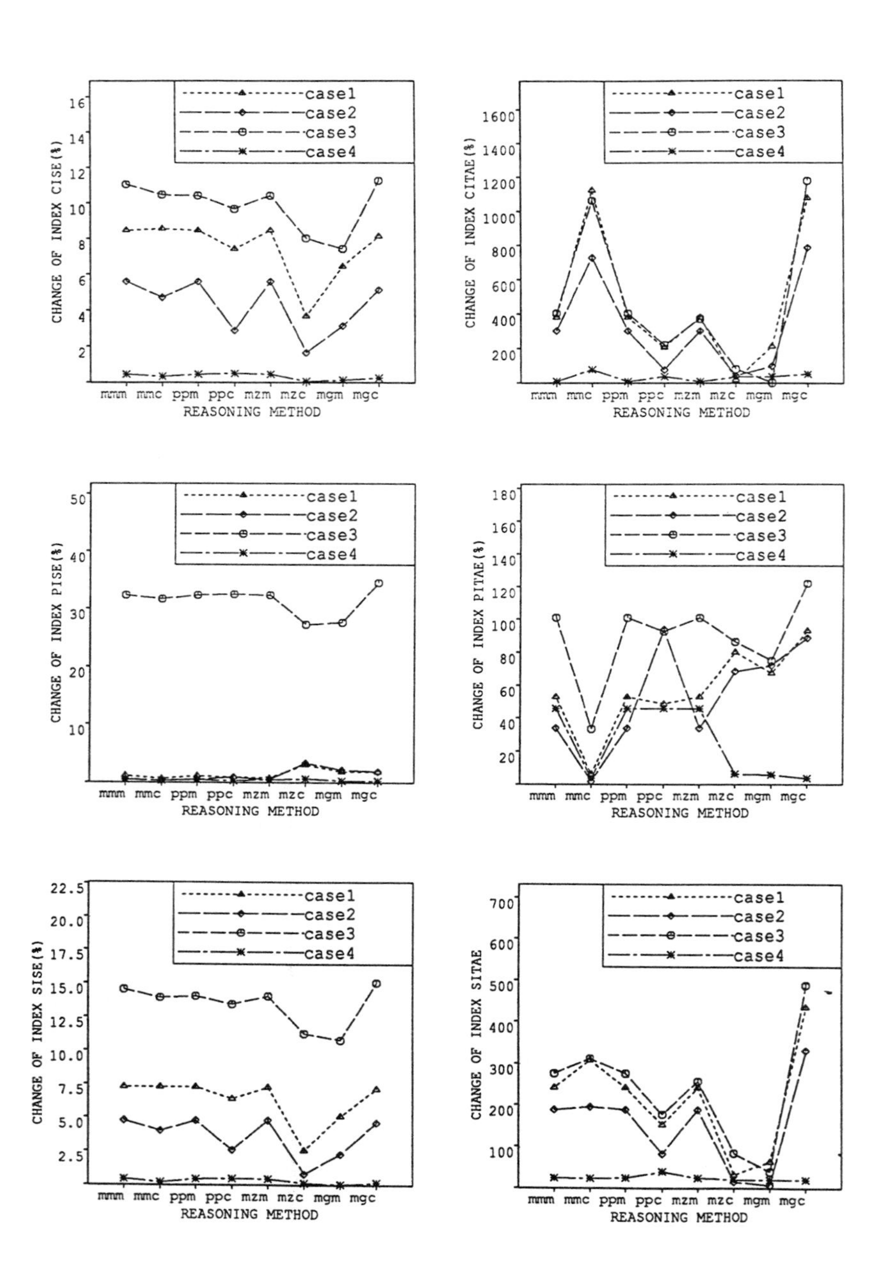


**Figure 3.17** Robustness to change in the compensator factors

are responsible for the poor performance. In contrast, the 25% changes in $\lambda_{12}$ and $\lambda_{21}$ (case 4) produced only a small variation in the indices and the responses were acceptable. It appears reasonable to conclude that the introduction of the compensator does eliminate the interactive effects effectively, and therefore the control performance can be improved by this simple scheme. In addition, the performance of the system is not affected greatly when a mild change is made in the factors, which indicates that useful robustness properties are possessed by the compensator.

## 3.4 Summary

The simulation results have demonstrated the feasibility of the proposed multivariable fuzzy controller when it is applied to the different processes under consideration. In our case, both fast and slow human blood pressure dynamic processes have been controlled successfully. Comparative simulation results have revealed that the selection of the reasoning algorithms should not be too crucial. However, from the robustness and simplicity point of view, it seems that MMM, PPM and PPC algorithms are good candidates to be selected for real-time control applications.

The decentralized control strategy simplifies the design procedures and overcomes the difficulty in constructing the rule-base. In this study, a commonsense rule-base has been used to control two different processes. The interactive effects between the variables are partly handled by introducing a simple compensation scheme, whose realization is based on the process steady-state gains. The simulation results showed that the scheme is necessary and efficient in compensating the cross-coupling effects and, perhaps more important, that the compensator possesses useful robustness properties against variation in the gains.

A point worth mentioning is the applicability of the fuzzy controller when dealing with nonlinear systems. For a moderately nonlinear plant, the error-based fuzzy controller as described in this chapter is appropriate. However, for severely nonlinear systems, it is necessary to tune some controller parameters to maintain good performance. One choice is to adjust the scaling factors in response to different set-point demands.

The proposed multivariable fuzzy controller is applicable only if the steady-state gains of the controlled process are available, although they do not need to be known accurately. As discussed previously, the gains may be obtained through several methods including theoretical and experimental approaches. If this is not the case, a fully cross-coupling fuzzy controller must be adopted which gives rise to the difficulty of building a rule-base consisting of rules with multiple-premises. This will be the main concern of the following chapters where the self-learning strategies are intensively investigated.

CHAPTER 4

# Constructing rule-bases by self-learning: system structure and learning algorithm

---

*A novel method capable of constructing rule-bases automatically via self-learning is presented. Three learning update laws are suggested and the convergence property of the learning algorithms is analysed in the sense of some defined norms.*

---

## 4.1 Introduction

An obvious and fundamental question arising from the real implementation of rule-based control systems (or more broadly expert control systems) is how a set of control rules can be derived. In general expert systems, this issue is termed as knowledge acquisition and is thought of as a bottleneck for building a realistic expert system. Likewise, the success and performance of rule-based control systems depend largely on the availability and the performance of the rule-base. When there are no experts or skilled operators available at hand to supply necessary knowledge or to be used as an identified model, it is necessary to construct the rule-base by directly operating the process being controlled. It is also desirable to refine and improve the rough rule-base derived from experts which may be incomplete, inconsistent or even partly incorrect, especially when the operating condition is changed.

This chapter together with the next, which also deals with self-learning, proposes a new methodology which can be used to construct rule-bases for multivariable fuzzy controllers via self-learning. The controlled object, the multivariable process, is assumed to be characterized by strong interaction between variables and pure time delays in control. The aim of the proposed system is to construct rule-bases which are independent and decoupled and can be used for several individual control loops. The basic idea of the proposed system may be simply stated as follows. By introducing a reference model and employing an iterative learning control scheme, the desired control actions are learned. At the same time, the rule-bases are formed by observing, recording and properly processing the learned actions which are used subsequently. The chapter focuses on the issues of the system structure and the learning schemes, whereas the next chapter involves the issues of rule-base formation and its application to a multivariable blood pressure control problem.

In Section 4.2, the basic structure of the proposed system is outlined. Beginning with a brief review of the learning control, Section 4.3 presents three learning algorithms with some comments about them. The convergence analysis and necessary proof of the learning process are given in Section 4.4.

## 4.2 Description of system structure

One of the most important and attractive characteristics of fuzzy controllers is that only little explicit knowledge regarding the controlled process is needed while designing the controller. Therefore, this feature should be preserved in developing any kind of fuzzy controller. Suppose that the process is a multivariable system with $n$ inputs and $n$ outputs subjected to step command signals. The process is not necessarily linear and possibly has pure time delay. The following requirements should be satisfied.

1. The control system should satisfy the desired performance in terms of transient and steady-state indices.
2. The required knowledge about the process should be kept as little as possible. Neither a fuzzy nor a nonfuzzy process model is needed.
3. The algorithm should be as simple as possible.

Following these requirements, we propose a rule-base construction system which is aimed at extracting control rules from the process of the iterative learning. This means that the correct control actions are progressively learned by operating the system repeatedly. Clearly, this process is similar to the learning process possessed by a human being. The block diagram of the proposed system is shown in Figure 4.1 in which the controlled process is assumed to have two inputs and two outputs. The overall system is composed of four functional modules: the reference model, the learning algorithm, the rule-base formation mechanism and the controlled process. The function of each component is outlined briefly as follows.

1. *Reference model*
   The reference model with inputs $r = [r_1,r_2]^T$ and outputs $y_d = [y_{d1},y_{d2}]^T$ is used to designate the desired performance. Notice that although the process may have strong interaction effects between the two inputs, the desired performance for the two output variables can be represented by two independent models, each of which has an appropriately chosen input. In other words, the reference model is

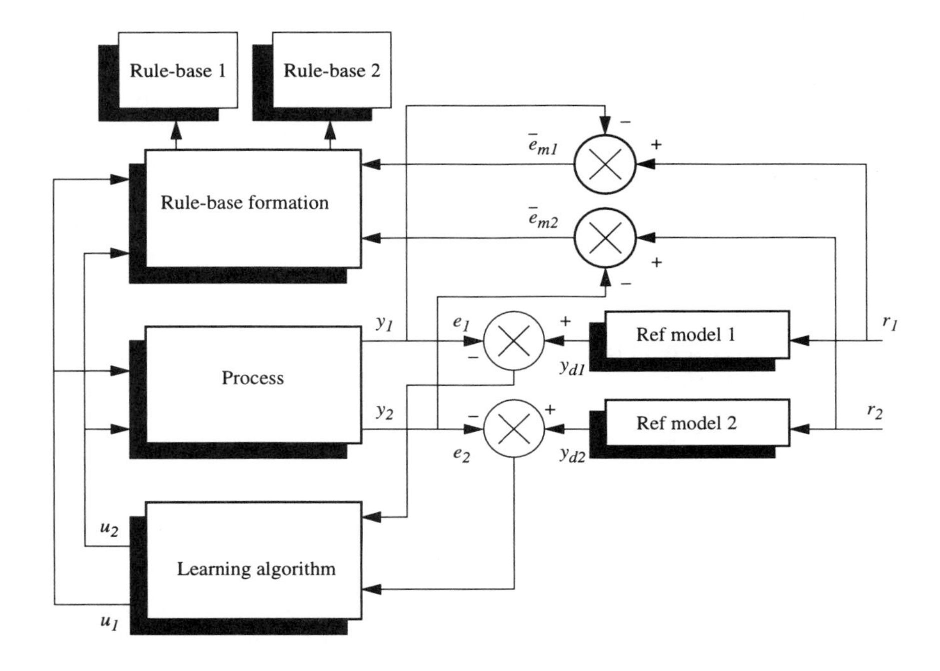


**Figure 4.1** Overall system structure for constructing rule-bases

non interacting. The model can be a low-order linear one described by either a transfer function or state equation. For example, the transfer functions of the first and second-order systems with pure time delay are given by

$$H_r(s) = \frac{K_r e^{-\tau_r s}}{T_r s+1} \tag{4.1a}$$

and

$$H_r(s) = \frac{\omega_{nr}^2 e^{-\tau_r s}}{s^2+2\xi_r \omega_{nr} s+\omega_{nr}^2} \tag{4.1b}$$

The parameters in the model are determined by the desired time domain indices such as overshoot, settling time and steady-state error, or alternatively by the desired pole position in the $s$-plane. Thus the outputs of the models represent the desired outputs.

2. *Learning algorithm*
   The main function of the learning algorithm is to obtain the correct control signal $u_d = [u_1,u_2]^T$ corresponding to the desired output $y_d$. The error $e = [e_1,e_2]^T$,

called the learning error and defined as the difference between the desired response $y_d$ and measured process output $y$, i.e. $e = y_d - y$, is used as a learning signal. It is expected that $e$ will asymptotically approach to zero or to a predefined small region with increase in the trial number. It is worth noting that this asymptotical property is along the iteration direction not along the time direction. More specifically, we require that $||e_k(\mathrm{t})|| \rightarrow 0$ or $||e_k(t)|| < \gamma$ in $t \in [0,T]$ as $k \rightarrow \infty$, where $k$ denotes the iteration number, $\gamma > 0$ is a predefined error tolerance and $|| \cdot ||$ stands for norm. Thus convergence occurs uniformly in the time interval of interest $[0,T]$. Obviously, the desired control $u_d$ is learned at this time. This scheme is different from traditional adaptive control strategies, for instance Model Reference Adaptive Control in which $||e_k(t)|| \rightarrow 0$ asymptotically as $t \rightarrow \infty$. In addition to learning $u_d$, the learning algorithm will learn appropriate scaling factors that are subsequently used for fuzzy control. The learning algorithm will be presented in detail in the next section.

3. *Rule-base formation*

   Parallel to learning, the rule-base formation mechanism at each trial creates two sets of control rules for two input–output pairs. Notice that although the multivariable process may possess strong interaction between the two inputs, the control rules can be set up independently since interaction effects are suitably and implicitly embedded into the learned $u_d$. Each rule, relating the measured control error $\bar{e}_m$ defined as $\bar{e}_m = r - y$ and change-in-error $\bar{c}_m$ to control action $u$, has the form "IF $E = \bar{e}$ and $C = \bar{c}$ THEN $U = u$", where $\bar{e}$ and $\bar{c}$ belonging to respective universes of discourse are mapped from measured values $\bar{e}_m$ and $\bar{c}_m$ with corresponding scaling factors. It is worth noting that the rules are purely value-based in the sense that there is no time relationship embedded in the rule variables. The rule-bases are memoryless, meaning that they are updated at each trial and only rely on the values of $\bar{e}$, $\bar{c}$ and $u$ obtained at the present trial. The methods for building the rule-base will be detailed in the next chapter.

*Remark:* It is worth discussing the prior knowledge needed to implement the proposed scheme. Intuitively, this lies mainly in the selection of the reference model besides the knowledge possibly required by the learning algorithm. It would be unrealistic to select the model parameters if one knows nothing about the controlled process. For example, we cannot demand instant rise time of the controlled output following a step command due to the presence of the process inertia and the time delay and possibly the magnitude limitation of the process input. In this sense, the selection of the reference model is dependent, to some extent, on an understanding of the process characteristics, though not fully or quantitatively.

## 4.3 Proposed learning algorithm

### 4.3.1 Overview

Recently, a considerable amount of research has been devoted to the development of a novel control method, called iterative learning control, and its application to robotic tracking. As its name implies, the correct control action is progressively learned and hence the desired performance is progressively achieved by repeated trial in such a way that the modification of the present control is based on the error information obtained during previous trials. The most attractive features of the method include:

1. Tight control can be obtained in the time interval of interest.
2. Less calculation is needed.
3. More importantly, less *a priori* knowledge about the process is required.

The concept and algorithm were formally presented by Arimoto and his colleagues (Arimoto *et al.* 1984, 1985, 1986) and was intensively treated from the theoretical viewpoint by Arimoto (1990). A learning algorithm called a PID-type update law was developed, and the learning convergence property for the robotic dynamic system was proved. In addition, robustness and boundedness of the learning algorithm were analysed. Following Arimoto's work, some similar concepts and algorithms have been proposed. For example, Bien and Hah (1989) presented a higher-order iterative learning control algorithm in which multiple past-history data pairs were used to update the present control. Geng *et al.* (1990) treated the learning control system as a two-dimensional system so that the analysis can be performed by means of established 2-D theory. Poter and Mohamed (1991) presented an improved learning algorithm which can be applied to the control of partially irregular plants. Hwang *et al.* (1991) described a learning algorithm for a class of linear discrete-time dynamic system with unknown but periodic parameters.

It is worth mentioning that there also exists a control scheme called repetitive control analogous to the learning control in which the repetitive operation is done continuously. The work has been reported in, for instance, Hara *et al.* (1988), Sadegh *et al.* (1990) and Messner *et al.* (1991).

### 4.3.2 Proposed learning algorithm

All the work reported above is primarily limited to robotic tracking applications. The analysis model used to prove convergence is either ordinary differential equations describing the robotic dynamics or state-space equation representation for a controlled process. Here, as mentioned earlier, we are interested in the problem of how a set of control actions can be learned subject to step command signals for a multivariable system with time delays in control.

Consider first a two-input two-output linear system described by

$$\begin{bmatrix} Y_1(s) \\ Y_2(s) \end{bmatrix} = \begin{bmatrix} H_{11}(s) & H_{12}(s) \\ H_{21}(s) & H_{22}(s) \end{bmatrix} \begin{bmatrix} e^{-\tau_1 s} & 0 \\ 0 & e^{-\tau_2 s} \end{bmatrix} \begin{bmatrix} U_1(s) \\ U_2(s) \end{bmatrix} \tag{4.2}$$

where $Y(s) = [Y_1(s), Y_2(s)]^T$ and $U(s) = [U_1(s), U_2(s)]^T$ are the Laplace transforms of the controlled output $y(t) \in R^2$ and the control input $u(t) \in R^2$ respectively.

Let $y_{d1} = [y_{d1}, y_{d2}]^T$ be the desired output, where $y_{di} = L^{-1}[H_{ri}(s)R_i(s)]$ for $i$ = 1, 2. Define the learning error signal $e(t)$ as $e(t) = y_d(t) - y(t)$. The objective of the learning control is to determine the control input $u(t)$ by repetitive trial such that the error asymptotically tends to zero, or a prespecified small value, in the time interval of interest. We propose the following learning algorithms:

1. Error correction algorithm

$$\begin{cases} u_{1,k+1}(t) = u_{1,k}(t) + g_1 e_{1,k}(t+\lambda_1) \\ u_{2,k+1}(t) = u_{2,k}(t) + g_2 e_{2,k}(t+\lambda_2) \end{cases} \tag{4.3}$$

2. Cross-error correction algorithm

$$\begin{cases} u_{1,k+1}(t) = u_{1,k}(t) + g_{11} e_{1,k}(t+\lambda_1) + g_{12} e_{2,k}(t+\lambda_2) \\ u_{2,k+1}(t) = u_{2,k}(t) + g_{22} e_{2,k}(t+\lambda_2) + g_{21} e_{1,k}(t+\lambda_1) \end{cases} \tag{4.4}$$

3. Error and derivative correction algorithm

$$\begin{cases} u_{1,k+1}(t) = u_{1,k}(t) + p_1 e_{1,k}(t+\lambda_1) + q_1 \dot{e}_{1,k}(t+\lambda_1) \\ u_{2,k+1}(t) = u_{2,k}(t) + p_2 e_{2,k}(t+\lambda_2) + q_2 \dot{e}_{2,k}(t+\lambda_2) \end{cases} \tag{4.5}$$

where $k$ denotes iteration number, $\lambda_i$, $i$ = 1, 2 is time advance and $g$, $p$, $q$ are learning gains.

If we assume that $\lambda_1 = \lambda_2 = \lambda$, then the above algorithms may be represented in a compact form given by

$$u_{k+1}(t) = u_k(t) + Pe_k(t+\lambda) + Q\dot{e}_k(t+\lambda) \tag{4.6}$$

where $P$ and $Q$ are learning matrices. It is clear that equations (4.3), (4.4), and (4.5) can be derived from (4.6) if $P$ and $Q$ are suitably chosen. However, we keep the algorithms separated in order to stress the loop effects upon the learning process and to show how the algorithm can be intuitively derived from common control knowledge as discussed later.

The discrete versions of the above algorithms are given in a compact form by

$$u_{k+1}(n) = u_k(n) + Pe_k(n+\mu) + Qc_k(n+\mu) \tag{4.7}$$

where $n$ denotes the nth sampling instant, $\mu$ is an integer representing the advance sampling number and $c_k(n)$ is change-in-error defined by $c_k(n) = e_k(n+1) - e_k(n)$. When $P = \text{diag}\{p_1, p_2\}$ and $Q = 0$, the discrete version corresponding to equation (4.3) is obtained. Likewise Q = 0 corresponds to that of equation (4.4) and $P = \text{diag}\{p_1, p_2\}$ with $Q = \text{diag}\{q_1, q_2\}$ corresponds to that of equation (4.5).

Some comments are in order. First, it is noted that the error between the step command signal $r(t)$ and the controlled output $y(t)$ cannot be used as a learning basis because such a learning objective (step output) is clearly unrealistic. Therefore, a reference model is introduced which specifies an achievable performance one would like to attain. The error between the output of the reference model and the controlled output is used as a learning signal and this error will approach zero with the increase of the iteration number $k$.

Second, the control update at the $(k+1)$th trial in all the algorithms is based on the information obtained at the $k$th trial, namely, the previous control action and the error. However, there are some differences in the three algorithms resulting from the considerations of the effects of loop interaction. As mentioned in the previous section, the ultimate objective of the proposed system is to construct two sets of control rules for two control loops. This implies that two input–output pairs are properly selected and the system appears to be fully decoupled. Accordingly, the reference model is designated separately for two input–output pairs. Suppose that $u_1/y_1$ and $u_2/y_2$ are suitably paired. Algorithm 1 is a decoupling update law in the sense that the learning in each control loop utilizes only the error information created by its own loop. Thus like the structure of the reference model, the learning mechanism is separated. It should be noted that this separation does not suggest that the effects of the loop interaction are neglected completely. In fact, this effect is implicitly embedded in the corresponding error signal. The loop effects are treated explicitly in algorithm 2 leading to the fact that the learning gain matrix is no longer diagonal. The third update law is again a separated algorithm for two loops. However, the error rate is taken into account in this case in the hope that the learning performance would be improved compared with the first algorithm since more information is used.

Third, we notice that the update learning law at the $n$th instant is based on the error that occurred at the $(n+\mu)$th instant. This results from the consideration that the control action at the $n$th instant can only primarily make the error at the $(n+\mu)$th instant small due to the time delay in the process. Note that $e_k(n+\mu)$ is available at the $(k+1)$th interaction at the $n$th time instant. It is the iterative property that makes the problem of output prediction simple which has been a main concern in the traditional controller design for the delay system. More complicated algorithms can be constructed if the errors beyond the $\mu$th instant are considered. This may be given by

$$u_{k+1} = u_k + \sum_{i=1}^{m} G_i e_k(\mu+i) \tag{4.8}$$

where $G_i = \text{diag}\,\{g_{i1}, g_{i2}\}$, $i = 1, m$.

However, the difficulty arising from the implementation of the above algorithm lies in the determination of matrices $G_i$.

Finally, it is interesting to compare the learning algorithm with the well-known Model Reference Adaptive Control (MRAC) (Astrom and Wittenmark 1989). In MRAC, the objects to be adjusted are some parameters, for instance feedforward and/or feedback gains, whereas in the learning system, the control input is directly updated. MRAC is designed to force the error between model output and the process output asymptotically to zero with increasing time by regulating the adaptive parameters. In contrast, the learning system is intended to eliminate this error uniformly in a time interval of interest and this objective is asymptotically achieved with the increase of the iteration number. What is common to them is that the update law in both cases is proportional to the tracking error although the update direction is different. However, there is a distinct difference between them. Namely, the determination of the update law in MRAC is based on some well-established theories, for example the gradient theory, Lyapunov stability theory and passivity theory, whereas in the learning control this determination is less theory-based and is heuristic in nature. Consequently, less prior quantitative knowledge about the process is needed in the latter case. However, it is worth noting that the learning control laws in the learning system can also be derived from the derivation of a squared performance measure with respect to the control input rather than to the controller parameters (Togai and Yamano 1985). Although the obtained learning laws may be optimal in some sense, they are in fact difficult to apply since some exact quantitative knowledge about the process being controlled is required.

## 4.4 Convergence analysis

### 4.4.1 Error correction

Consider a system described by

$$Y(s) = H(s)e^{-hs}U(s) \tag{4.9}$$

where $Y(s) = L[y(t)]$, $y \in R^m$, $U(s) = L[u(t)]$, $u \in R^m$, $H(s)$ is a $m \times m$ transfer matrix and $h$ denotes the time delay in control which is assumed to be the same for all inputs. Suppose that $H(s)$ is a strictly proper and stable rational transfer matrix, then the system may be represented by

$$\begin{aligned} \dot{x}(t) &= Ax(t) + Bu(t-h) \\ y(t) &= Cx(t) \end{aligned} \tag{4.10}$$

where $x(t) \in R^n$ is a state vector and $A$, $B$, and $C$ are constant matrices with appropriate dimensions.

As already mentioned, the aim of the control is to force the controlled output $y(t)$ to follow the step command signal $r(t)$ with the desired performance specified by the reference model. In other words, $y_d(t)$ and $u_d(t)$ are the desired output and input. Because $y_d(t)$ and $u_d(t)$ must reach a steady-state after a transient period $T_t$, it is sufficient to learn the $u_d(t)$ in the interval $[0,T-h]$, where $T > T_t > h$. In what follows, $t$ will be restricted to $[0,T]$. Accordingly, $u_d(t)$ and $u_{k(t)}$ in $t \in [0,T]$ are defined by

$$u_d(t) = \begin{cases} u_d(t) & t \in [0,T-h] \\ u_d(T-h) & t \in [T-h,T] \end{cases}$$

$$u_k(t) = \begin{cases} u_k(t) & t \in [0,T-h] \\ u_k(T-h) & t \in [T-h,T] \end{cases} \tag{4.11}$$

From equation (4.10), we obtain

$$\begin{aligned} y_k(t) &= C\exp[At]x_k(0) + \int_0^t Cexp\,[A(t-\tau)]Bu_k(\tau-h)\mathrm{d}\tau \\ &= g(t) + \int_0^t Ch(t-\tau)Bu_k(\tau-h)\mathrm{d}\tau \end{aligned} \tag{4.12}$$

where $g(t) = C\exp[At]x_k(0)$ and $h(t) = C\ \exp[A(t)]\ B = L^{-1}\ [H(s)]$.

It should be noted that equations (4.10) and (4.12) are applicable only in $t \in [h,T]$. A complete definition of $y_k(t)$ in $[0,T]$ is given below.

$$y_k(t) = \begin{cases} g(t) & t \in [0,h] \\ g(t) + \int_0^t Ch(t-\tau)Bu_k(\tau-h)\mathrm{d}\tau & t \in [h,T] \end{cases} \tag{4.13}$$

The problem can be formulated as follows: to find a control vector $u_{d(t)}$, $t \in [0,T-h]$ so that the controlled output $y_k(t)$ approaches $y_d(t)$ uniformly in $t \in [h,T]$ as $k$ approaches infinity.

*Definition 1:* The $L_2$-norm of a vector-valued function $v(t) \in R^m$, $t \in [0,T]$, is defined by

$$||v(t)|| = [\int_0^T v^T(t)v(t)\mathrm{d}t]^{1/2} \tag{4.14a}$$

*Definition 2:* The induced norm of $m \times m$ matrix $G(j\omega)$ corresponding to the $L_2$-norm is defined by

$$||G(j\omega)|| = \max_i \lambda_i [G(j\omega)^* G(j\omega)] = \bar{\sigma}[G(j\omega)] \tag{4.14b}$$

where $\lambda_i [M]$ denotes the $i$th eigenvalue of the Hermitian matrix $M$, * denotes conjugate transpose and $\bar{\sigma}\,[M]$ denotes the largest singular value of the $M$.

*Definition 3:* The infinity norm of $G(j\omega)$ is defined by

$$||G(j\omega)||_\infty = \underset{\omega}{Sup}\ \bar{\sigma}[G(j\omega)] \tag{4.15a}$$

*Definition 4:* A Linear Time-Invariant (LTI) system is said to be strictly positive if, for any $T > 0$ and any input $u(t)$, $t \in [0,T]$, the following inequality is satisfied with some constant $\alpha > 0$

$$\int_0^T\int_0^t u^T(t)h(t-\tau)u(\tau)d\tau dt \geq \alpha||u(t)||^2 \tag{4.15b}$$

Now we have

*Theorem 1a:* Let the LTI system with delay $h$ in control be given by equation (4.10) and the learning control law be given by

$$u_{k+1}(t) = u_k(t) + Ge_k(t+h) \qquad t \in [0,\ T-h] \tag{4.16}$$

where $G \in R^{m \times m}$ and $e_k(t) = y_d(t) - y_k(t)$, $t \in [h,\ T]$.
Assume that

1. $g_k(t) = g(t) \qquad t \in [0,h] \qquad k = 1, 2, \cdots$
2. $G$ is chosen in such a way that $H(t) = h(t)\ G$ is strictly positive and the condition $\beta < 2\ \alpha$ (as defined later) is satisfied.

Then, $y_k(t) \rightarrow y_d(t)$, $t \in [h,\ T]$, as $k \rightarrow \infty$.

*Proof:* We follow an approach used by Arimoto *et al.* (1985). By the definition of the error, it follows that

$$\begin{aligned} e_{k+1}(t) - e_k(t) &= [y_d(t) - y_{k+1}(t)] - [y_d(t) - y_k(t)] \\ &= y_k(t) - y_{k+1}(t) \end{aligned} \tag{4.17}$$

Substituting equation (4.12) into (4.17), we obtain

$$e_{k+1}(t) - e_k(t) = \int_0^t h(t-\tau)[u_k(\tau-h) - u_{k+1}(\tau-h)]d\tau \tag{4.18}$$

Here assumption 1 is used. From equation (4.16), it follows that

$$e_{k+1}(t) - e_k(t) = -\int_0^t h(t-\tau)Ge_k(\tau)\mathrm{d}\tau$$

Thus,

$$e_{k+1}(t) = e_k(t) - \int_0^t H(t-\tau)e_k(\tau)\mathrm{d}\tau \tag{4.19}$$

where $H(t-\tau) = h(t-\tau)G$.

Computing the norm of both sides of equation (4.19), we have

$$\|e_{k+1}(t)\|^2 = \|e_k(t)\|^2 + \|W_k(t)\|^2 - 2\cdot\int_0^t e_k(t)^T W_k(t)\mathrm{d}t \tag{4.20}$$

where

$$W_k(t) = \int_0^t H(t-\tau)e_k(\tau)\mathrm{d}\tau \tag{4.21}$$

According to Parseval's equality in relation to the Fourier transform in function space $L_2[0,T]$ (Desoer and Vidyasagar 1975), we have

$$\begin{aligned}\|W_k(t)\|^2 &\leq [\sup_\omega \|\hat{H}(\mathrm{j}\omega)\|]^2 \|e_k(t)\|^2 \\ &= \|\hat{H}(\mathrm{j}\omega)\|_\infty^2 \|e_k(t)\|^2 \\ &= \beta\|e_k(t)\|^2\end{aligned} \tag{4.22}$$

where $\hat{H}(\mathrm{j}\omega)$ is the Fourier transform matrix of $H(t)$ and $\beta = \|\hat{H}(\mathrm{j}\omega)\|_\infty^2$.

Noting the strictly positive assumption and equation (4.22), we obtain

$$\|e_{k+1}(t)\|^2 \leq (1 + \beta - 2\alpha)\|e_k(t)\|^2 \tag{4.23}$$

By choosing $G$ so that $\rho^2 = (1 + \beta - 2\alpha) < 1$, i.e. $\beta < 2\alpha$ we have

$$\|e_{k+1}(t)\| \leq \rho\|e_k(t)\| \qquad 0 \leq \rho < 1 \tag{4.24}$$

which implies

$$\|e_{k+1}(t)\| \leq \rho^k\|e_0(t)\| \tag{4.25}$$

This means $e_k(t)\to 0$ as $k\to\infty$ and hence $y_k(t)\to y_d(t)$ as $k\to\infty$, where $t \in [h,T]$.

Theorem 1a can be stated in frequency terms as follows:

*Theorem 1b:* Suppose the delayed system and learning control law are the same as in Theorem 1a. If $H(s)$ has no zeros on the imaginary axis and learning gain matrix $G$ is so chosen that $\|I-H(\mathrm{j}\omega)G\|_\infty < 1$, then $y_k(t)\to y_d(t)$ as

$k \to \infty$, where $t \in [h, T]$.

*Proof:* From equation (4.9) and the definition of the error in the $s$-domain, it follows that

$$E_{k+1}(s) - E_k(s) = Y_k(s) - Y_{k+1}(s)$$
$$= H(s)e^{-hs}[U_k(s) - U_{k+1}(s)] \tag{4.26}$$

Substituting the learning law in the $s$-domain into equation (4.26), we have

$$E_{k+1}(s) - E_k(s) = -H(s)e^{-hs} \cdot GE_k(s)e^{hs}$$
$$= -H(s)GE_k(s) \tag{4.27}$$

Hence,

$$E_{k+1}(s) = E_k(s) - H(s)GE_k(s)$$
$$= [I - H(s)G]E_k(s) \tag{4.28}$$

Replacing $s$ by $j\omega$ in equation (4.28), we obtain

$$E_{k+1}(j\omega) = [I - H(j\omega)G]E_k(j\omega) \tag{4.29}$$

Taking the norm for both sides of equation (4.29), we have

$$||E_{k+1}(j\omega)|| \le ||I - H(j\omega)G|| \; ||E_k(j\omega)||$$
$$\le ||I - H(j\omega)G||_\infty \; ||E_k(j\omega)|| \tag{4.30}$$

Denote $||I - H(j\omega)G||_\infty$ as $\rho$. It follows that

$$||E_{k+1}(j\omega)|| \le \rho^k \cdot ||E_0(j\omega)|| \tag{4.31}$$

Since $\rho < 1$ by assumption, we have

$$||E_k(j\omega)|| \to 0 \qquad as \;\; k \to \infty$$

which implies $||e_k(t)|| \to 0$ and hence $y_k(t) \to y_d(t)$.

A special case of Theorem 1 is that we wish to set the learning gain matrix $G$ to be diagonal implying that there is no cross-error contributing to the learning. We have the following result:

*Theorem 2:* Suppose that the delayed system and the learning law are the same as given in Theorem 1a with the assumption that

1. $g_k(t) = g(t) \qquad t \in [0, h] \qquad k = 1, 2, \cdots$
2. LTI system is strictly positive.
3. $G = \gamma I_{m \times m}$, where $\gamma$ is a sufficiently small positive number.
   Then, $y_k(t) \to y_d(t)$ as $k \to \infty$, where $t \in [h, T]$.

The proof of the Theorem can be found in Arimoto *et al.* (1985) but with the difference that no control delay was considered.

### 4.4.2 Error and derivative correction

When the error rate is taken into account, we have the following results:

1. Continuous case

   As pointed out by Arimoto *et al.* (1985), a more general class of the system can be proved to be convergent under the learning law in which both error and derivative of error are considered. Suppose that the system with delay in control is given by

$$\begin{cases} \dot{x}(t) = f(t, x(t)) + Bu(t-h) \\ y(t) = Cx(t) \end{cases} \tag{4.32}$$

   where $x(t) \in R^n$ is a state vector-valued function, $f \in R^n$ is a nonlinear vector-valued function, $B \in R^{n \times m}$ and $C \in R^{m \times n}$ are constant matrices. Also we have the following definitions:

*Definition 5:* The following norms are defined

$$\begin{aligned} |x(t)|_\infty &= \max_i |x_i(t)| \qquad \text{for } x(t) \in R^n \\ |A|_\infty &= \max_i \sum_{j=1}^{n} |a_{ij}| \qquad \text{for } A \in R^{n \times n} \\ \|e(t)\|_\lambda &= \mathop{Sup}_{t \in [0,T]} [e^{-\lambda t} \cdot |e(t)|_\infty] \quad \text{for } e(t) \in R^n \end{aligned} \tag{4.33}$$

where $e(t)$ is defined over $t \in [0,T]$ and $\lambda$ is a positive number.

*Definition 6:* A function $f(t, x)$ is said to be Lipschitz in $x$ if there exists a constant $l \geq 0$ such that

$$|f(t, x_1) - f(t, x_2)|_\infty \leq l \cdot |x_1 - x_2|_\infty \tag{4.34}$$

for any pair $(x_1, x_2)$ in $R^n \times R^n$, and any $t \in [0, T]$.

Now Theorem 3a is stated as follows.

*Theorem 3a:* Let the system with delay in control be given by (4.32) and the learning control law be given by

$$u_{k+1}(t) = u_k(t) + Pe_k(t+h) + Q\dot{e}_k(t+h) \tag{4.35}$$

where $P$, $Q \in R^{m \times m}$ and $e_k(t) = y_d(t) - y_k(t)$.

Assume that

**(a)** $x_k(t) = y_d(t)$, $t \in [0, h]$, for $k = 1, 2, \ldots$

**(b)** $f$ is Lipschitz.

**(c)** $u_0(t) \in C^1[0, T-h]$ and $y_d(t) \in C^1[h, T]$.

**(d)** $|I-QCB|_\infty < 1$.

Then $y_k(t) \to y_d(t)$ as $k \to \infty$ uniformly in $t \in [h, T]$.

*Proof:* Suppose that there exists a control input such that

$$\dot{x}_d(t) = f(t, x_d(t)) + Bu_d(t-h) \tag{4.36}$$

It follows from equations (4.32), (4.35) and (4.36) that

$$\begin{aligned} u_d(t) - u_{k+1}(t) &= u_d(t) - u_k(t) - Pe_k(t+h) - Q\dot{e}_k(t+h) \\ &= u_d(t) - u_k(t) - PC[x_d(t+h) - x_k(t+h)] - QC[\dot{x}_d(t+h) \\ &\quad - \dot{x}_k(t+h)] \\ &= u_d(t) - u_k(t) - PC[x_d(t+h) - x_k(t+h)] \\ &\quad - QC[f(t+h, x_d(t+h)) + Bu_d(t) - f(t+h, x_k(t+h)) - Bu_k(t)] \\ &= u_d(t) - u_k(t) - PC[x_d(t+h) - x_k(t+h)] - QCB[u_d(t) - u_k(t)] \\ &\quad - QC[f(t+h, x_d(t+h)) - f(t+h, x_k(t+h))] \end{aligned} \tag{4.37}$$

Taking the norm for both sides of equation (4.37) and using the Lipschitz condition, we have

$$\begin{aligned} |u_d(t) - u_{k+1}(t)|_\infty &\le |I-QCB|_\infty |u_d(t) - u_k(t)|_\infty + [|PC|_\infty + |QC|_\infty \cdot l] \cdot |x_d(t+h) \\ &\quad - x_k(t+h)|_\infty \\ &= \alpha_1 \cdot |u_d(t) - u_k(t)|_\infty + \alpha_2 \cdot |x_d(t+h) - x_k(t+h)|_\infty \end{aligned} \tag{4.38}$$

where $\alpha_1 = |I-QCB|_\infty$ and $\alpha_2 = |PC|_\infty + |QC|_\infty \cdot l$

Because $x_k(h) = x_d(h)$ for all $k$, we obtain

$$\begin{aligned} |x_d(t+h) - x_k(t+h)|_\infty &= \left| \int_0^t [\dot{x}_d(\tau+h) - \dot{x}_k(\tau+h)] d\tau \right|_\infty \\ &= \left| \int_0^t [f(\tau+h, x_d(\tau+h)) + Bu_d(\tau) - f(\tau+h, x_k(\tau+h)) \right. \\ &\quad \left. - Bu_k(\tau)] d\tau \right|_\infty \\ &= \left| \int_0^t \{[f(\tau+h, x_d(\tau+h)) - f(\tau+h, x_k(\tau+h))] \right. \end{aligned}$$

$$+ B\cdot[u_d(\tau)-u_k(\tau)]\}d\tau|_\infty$$

$$\le \int_0^t [l\cdot|x_d(\tau+h)-x_k(\tau+h)|_\infty$$

$$+ |B|_\infty\cdot|u_d(\tau)-u_k(\tau)|_\infty]d\tau \qquad (4.39)$$

By applying the *Bellman Gronwall–Lemma* , we have

$$|x_d(t+h)-x_k(t+h)|_\infty \le \int_0^t |B|_\infty\cdot|u_d(\tau)-u_k(\tau)|_\infty\, e^{l(t-\tau-h)}\, d\tau \qquad (4.40)$$

Substituting equation (4.40) into (4.38), we obtain

$$|u_d(t)-u_{k+1}(t)|_\infty \le \alpha_1|u_d(t)-u_k(t)|_\infty$$

$$+ \alpha_2\cdot|B|_\infty\cdot e^{-lh}\int_0^t |u_d(\tau)-u_k(\tau)|_\infty\, e^{l(t-\tau)}\, d\tau \qquad (4.41)$$

Denoting $\alpha_2\cdot|B|_\infty e^{-lh}$ as $\alpha_3$ and multiplying both side of equation (4.41) by $e^{-\lambda t}$ with $\lambda > 0$, we have

$$e^{-\lambda t}\, |u_d(t)-u_{k+1}(t)|_\infty \le \alpha_1 e^{-\lambda t}|u_d(t)-u_k(t)|_\infty$$

$$+ \alpha_3\int_0^t e^{-\lambda\tau}\, |u_d(\tau)-u_k(\tau)|_\infty\, e^{(l-\lambda)(t-\tau)}\, d\tau \qquad (4.42)$$

From the definition of the norm $||\cdot||_\lambda$, it follows that

$$||u_d(t)-u_{k+1}(t)||_\lambda \le \alpha_1\, ||u_d(t)-u_k(t)||_\lambda$$

$$+ \frac{1-e^{(l-\lambda)T}}{\lambda-l}\cdot \alpha_3\, ||u_d(t)-u_k(t)||_\lambda$$

$$= [\alpha_1 + \frac{1-e^{(l-\lambda)T}}{\lambda-l}\cdot \alpha_3]\, ||u_d(t)-u_k(t)||_\lambda$$

$$= \rho\, ||u_d(t)-u_k(t)||_\lambda \qquad (4.43)$$

where $\rho = \alpha_1 + \dfrac{1-e^{(l-\lambda)T}}{\lambda-l}\cdot\alpha_3$

Since $\alpha_1 < 1$ by assumption, it is possible to choose $\lambda$ sufficiently large so that $\rho < 1$. Thus, $||u_d(t)-u_k||_\lambda \to 0$ as $k\to\infty$. Noting that

$$\sup_{t\in[0,T]} |u_d(t)-u_k(t)|_\infty \le \sup_{t\in[0,T]} e^{\lambda(T-t)}\, |u_d(t)-u_k(t)|_\infty = e^{\lambda T}||u_d(t)-u_k||_\lambda$$

we have

$$\sup_{t\in[0,T]} |u_d(t)-u_k(t)|_\infty \to 0 \qquad as \quad k\to 0 \qquad (4.44)$$

which implies $u_k(t) \rightarrow u_d(t)$ for $t \in [0,T-h]$ and hence $y_k(t) \rightarrow y_d(t)$ for $t \in [h,T]$.

2. Discrete case

Although it is possible to obtain a convergence proof for general discrete systems by adopting the similar procedures used in the continuous case, here we will treat the learning system as a two-dimensional discrete system and hence confine ourselves to linear systems only. Consider a discrete system with time delay in control described by

$$\begin{aligned} x[(i+1)T_s] &= Ax[iT_s] + Bu[(i-h)T_s] \\ y[iT_s] &= Cx[iT_s] \end{aligned} \tag{4.45}$$

where $T_s$ is the sampling period, $h$ is the delay number and $A$, $B$ and $C$ are constant matrices with appropriate dimensions. Here, we assume that the $C$ is a full rank matrix. In what follows, the $T_s$ will be omitted for simplicity. The learning update law is given in equation (4.7) and is repeated as follows for convenience:

$$u_{k+1}(i) = u_k(i) + Pe_k(i+h) + Qc_k(i+h) \qquad i \in [0,N-h] \tag{4.46}$$

where $P$ and $Q$ are learning gain matrices, $k$ is iteration number, $[0,N]$ is the time interval of interest, $e_k(i) = y_d(i) - y_k(i)$ is the error between the desired output and the measured output and $c_k(i) = e_k(i+1) - e_k(i)$ is the change-in-error. We note that $e_k(i)$ and $c_k(i)$ are defined over $[h, N]$. We have the following result:

*Theorem 3b:* Let the system be described by equation (4.45) with learning law of equation (4.46). Assume the following conditions are satisfied:

(a) $y_k(i) = y_d(i)$ for $i \in [0,h-1]$; $k = 0, 1, 2, \cdots$

(b) Spectral radius of $A$ is less than 1, i.e. $\lambda[A] < 1$.

(c) $P$ and $Q$ are so chosen that

$$-CABQ + CBP - CBQ = 0$$
$$\lambda[I - CBQ] < 1$$

Then, $y_k(i) \rightarrow y_d(i)$, $i \in [h,N]$, as $k \rightarrow \infty$.

*Proof:*

From equations (4.45) and (4.46), it follows that

$$\begin{aligned} e_k(i+1) &= y_d(i+1) - y_k(i+1) \\ &= y_d(i+1) - C \cdot [Ax_k(i) + Bu_k(i-h)] \\ &= y_d(i+1) - [CAC^{-1}Cx_k(i) + CBu_k(i-h)] \\ &= y_d(i+1) - [CAC^{-1}y_k(i) + CBu_k(i-h)] \end{aligned}$$

$$= y_d(i+1) - [M(y_d(i) - e_k(i)) + CBu_k(i-h)] \tag{4.47}$$

where $M = CAC^{-1}$. From equation (4.47), we have

$$e_{k+1}(i+1) = y_d(i+1) - [M(y_d(i) - e_{k+1}(i)) + CBu_{k+1}(i-h)] \tag{4.48}$$

Subtracting equation (4.47) from (4.48), we obtain

$$e_{k+1}(i+1) - e_k(i+1) = M[e_{k+1}(i) - e_k(i)] - CB[u_{k+1}(i-h) - u_k(i-h)] \tag{4.49}$$

Substituting the learning law into equation (4.49), we have

$$\begin{aligned} e_{k+1}(i+1) - e_k(i+1) &= M[e_{k+1}(i) - e_k(i)] - CB[Pe_k(i) + Qc_k(i)] \\ &= M[e_{k+1}(i) - e_k(i)] - CB[Pe_k(i) + Q(e_k(i+1) - e_k(i))] \\ &= Me_{k+1}(i) - (M + CBP - CBQ)e_k(i) - \\ &\quad CBQe_k(i+1) \end{aligned} \tag{4.50}$$

Let $W = M + CBP - CBQ$. Then

$$e_{k+1}(i+1) = Me_{k+1}(i) - We_k(i) + (I - CBQ)e_k(i+1) \tag{4.51}$$

If the iteration number $k$ and the sampling number $i$ are viewed as two independent coordinates (Geng *et al.* 1990) with $k = 0, 1, 2, \cdots$ and $i = h, h+1, \cdots, N$, the error equation (4.51) in fact consists of four adjacent errors located at $(i,k)$, $(i,k+1)$, $(i+1,k)$ and $(i+1,k+1)$ respectively. In this respect, equation (4.51) can be regarded as a two-dimensional system (2-D) error equation.

Similarly to the method in Geng *et al.* (1990), we define

$$\xi_k(i) = e_{k+1}(i) - (I - CBQ)e_k(i) \tag{4.52}$$

Therefore, equation (4.51) can be rewritten as

$$\begin{aligned} \xi_k(i+1) &= Me_{k+1}(i) - We_k(i) \\ &= M[\xi_k(i) + (I - CBQ)e_k(i)] - We_k(i) \\ &= M\xi_k(i) + (M - MCBQ - W)e_k(i) \end{aligned} \tag{4.53}$$

Noting that $W = M + CBP - CBQ$ and $M = CAC^{-1}$, we have

$$\xi_k(i+1) = M\xi_k(i) + (-CABQ + CBP - CBQ)e_k(i) \tag{4.54}$$

Combining equations (4.52) and (4.54), we obtain the following so-called Rosser 2-D error model

$$\begin{bmatrix} \xi_k(i+1) \\ e_{k+1}(i) \end{bmatrix} = \begin{bmatrix} M & -CABQ+CBP-CBQ \\ I & I-CBQ \end{bmatrix} \begin{bmatrix} \xi_k(i) \\ e_k(i) \end{bmatrix} \tag{4.55}$$

Applying the stability theorem for the 2-D system (Geng *et al.* 1990) and noting the conditions b) and c), we can conclude that the 2-D error system is stable, which implies that $e_k(i) \to 0$ as $k \to \infty$ and hence $y_k(i) \to y_d(i)$ as $k \to \infty$.

To end this section, we make some remarks about the learning algorithms and the convergence analysis presented in the section.

*Remark 1*: There is no explicit mathematical model needed in the design of the learning law. What we are trying to do here is, by means of some standard mathematical models, to demonstrate how a simple algorithm can be used to learn the correct control action for a class of linear or nonlinear systems with time delay in the control.

*Remark 2*: We require that the control delay be known so that the learning law and the reference models can be implemented. It is assumed that $h$ is identical to every pair of $u_i/y_i$. In practice, there may exist different control delays. In that case, $h$ can be chosen to be different in the learning law with respect to the different control loops as given in equations (4.3), (4.4), and (4.5).

*Remark 3*: It is evident that the learning gain matrix plays an important role in guaranteeing the convergence and in determining the learning rate. Unfortunately, there is no general rule given for determining it. Therefore, trial and error is needed in practice.

## 4.5 Summary

A novel method for constructing rule-bases for the use of fuzzy controllers by self-learning is suggested. The controlled processes of interest are multivariable systems with strong interaction between variables and with pure time delays in control. The objective of the proposed system is to construct two separated and decoupled rule-bases for two control loops with some design constraints such as performance requirements subject to step command signals, little *a priori* knowledge about the process, and algorithm simplicity. We have presented a system structure consisting of four modules: the reference model, the learning algorithm, the rule-base formation mechanism and the process being controlled. By introducing the concept of learning errors, three learning algorithms have been proposed. The convergence analysis shows that the learning process using the proposed learning laws can be convergent in the sense of defined norms under some mild assumptions.

Finally, we notice that some issues regarding the learning algorithm have not been addressed here. For example, the robustness of the learning algorithm with respect to the time delay, the learning gains and the effect of noise-contamination on learning errors should be investigated. While realizing that there exist some difficulties in dealing with these issues theoretically, they can be studied in some detail by means of simulation and this is reported in the next chapter.

CHAPTER 5

# Constructing rule-bases by self-learning: rule-base formation and application

---

*A methodology for constructing a rule-base from learned data is described. By defining some performance measures the behaviour of the proposed system, in terms of its learning ability, reproducibility and robustness, is evaluated.*

---

## 5.1 Introduction

While the previous chapter was concerned primarily with the issue of system structure for constructing rule-bases and associated learning control laws, this chapter first of all focuses on developing a methodology to build the rule-base in an automatic manner based on the knowledge acquired during the learning stage, and then applies the proposed scheme to a problem of multivariable control of blood pressure.

The rule-base formation may be viewed as a matter of symbolic fuzzy modelling provided that the numerical non-fuzzy data sets regarding the inputs and outputs of the fuzzy controller are available. While there are no systematic tools available, to date, to handle this problem, it is observed that the method for building the rule-base is closely related to the reasoning mechanism employed, meaning that different inference strategies require different model structures. For example, a relation equation model is constructed in the work reported in Pedrycz (1985b). No rule-base is explicitly built due to the fact that the inference relies entirely on the relation equation.

It is very difficult to derive, from numerical data, a rule-base in the form of IF-THEN statements which is not only linguistically identical to the one acquired from expert's verbal description but which also satisfies some accuracy requirements. However, from an engineering control point of view, it is possible to construct a rule-base with the rules having less linguistic meaning but having an IF-THEN statement form. Based on this consideration, we present a relatively simple approach for extracting the rules from the recorded data. The rule-base is suitable for simplified fuzzy control algorithms. The rules are of the IF-THEN structure but no linguistic label is attached on the THEN part.

As an application and feasibility demonstration of the proposed system, the problem of multivariable control of blood pressure is studied by means of simulation. By defining several evaluation measures, learning ability subject to the change of performance specifications and the variation of process parameters is investigated. Reproducibility of the fuzzy controller utilizing the rule-base extracted from the learning control is also evaluated in terms of relative accuracy with respect to the learning results. In addition, the issue of learning speed, which is affected by learning gains, initial control and different learning laws, is explored.

The next section, Section 5.2, presents a simplified fuzzy control algorithm within which the acquired rule-base is employed. Section 5.3 is devoted to developing the method of extracting rules from recorded data. The results using this method for blood pressure control are presented in Section 5.4.

## 5.2 Fuzzy control algorithm

The reasoning scheme used here is a simplified algorithm proposed in Nie (1987, 1989) and is presented as follows. Since one of the benefits obtained from the learning stage is that the multivariable system is fully decoupled in the sense that two independent control loops can be built with the corresponding learned control action, it is sufficient to discuss only one control loop.

1. *Input and output*

   Assume that the input variables to the fuzzy controller are error $E$ and change-in-error $C$ which take their values on universes of discourse $\bar{E}$ and $\bar{C}$ respectively. We define $\bar{E}$ and $\bar{C}$ as finite, discrete and symmetric about zero, that is,

$$\bar{E} = \left\{-v_N, -v_{N-1}, \cdots, -v_1, 0, v_1, \cdots, v_{N-1}, v_N\right\} \qquad Card[\bar{E}] = 2N+1$$

$$\bar{C} = \left\{-w_M, -w_{M-1}, \cdots, -w_1, 0, w_1, \cdots, w_{M-1}, w_M\right\} \qquad Card[\bar{C}] = 2M+1 \tag{5.1}$$

   where $v_i$, $i \in [1, N]$ and $w_j \in [1, M]$ are positive and are assumed to be equally spaced on the real line.

   The output of the fuzzy controller is control action U which takes its value in $\bar{U}$. Here $\bar{U} = [u_{\min}, u_{\max}]$ is a closed and continuous interval with $u_{\min}$ and $u_{\max}$ being minimum and maximum admissible control values.

2. *Fuzzy subsets*

   Suppose that there are $N_f$ fuzzy sets $A_i$, $i \in [1, N_f]$ defined over $\bar{E}$ and $M_f$ fuzzy sets $B_j$, $j \in [1, M_f]$ defined over $\bar{C}$. $A_i$ and $B_j$ may represent some fuzzy concepts such as *positive-large* and *negative-small* etc. and their membership functions (MF) are defined by

$$A_i(\bar{e}) = \begin{cases} 1 - \left[ \dfrac{|\bar{e}_i^* - \bar{e}|}{\delta_i^e} \right] & \text{if } |\bar{e}_i^* - \bar{e}| \le \delta_i^e \\ 0 & \textit{otherwise} \end{cases} \tag{5.2}$$

$$B_j(\bar{c}) = \begin{cases} 1 - \left[ \dfrac{|\bar{c}_j^* - \bar{c}|}{\delta_j^c} \right] & \text{if } |\bar{c}_j^* - \bar{c}| \le \delta_j^c \\ 0 & \textit{otherwise} \end{cases} \tag{5.3}$$

where $\bar{e} \in \bar{E}$ and $\bar{c} \in \bar{C}$ denote the generic element of $\bar{E}$ and $\bar{C}$ respectively, $\bar{e}_i^* \in \bar{E}$, $\bar{c}_i^* \in \bar{C}$, $0 < \delta_i^e \le v_N$ and $0 < \delta_j^c \le w_M$. We see that $A_i(\bar{e})$ and $B_j(\bar{c})$ are of triangular form and possess some properties such as normality, interval monotonicity and symmetry about $\bar{e}_i^*$ and $\bar{c}_j^*$. Note that if $\bar{e}_i^*(\bar{c}_j^*)$ takes an extreme value in $\bar{E}(\bar{C})$, the corresponding MF is no longer symmetric and has a half triangular form.

From equations (5.2) and (5.3), we see that the MF of a fuzzy set is characterized by $\bar{e}_i^*(\bar{c}_j^*)$ and $\delta_i^e(\delta_j^c)$. More specifically, $\bar{e}_i^*(\bar{c}_j^*)$ represents the central value of the corresponding support set at which maximum grade 1 is assigned and $\delta_i^e(\delta_j^c)$ designates the half width of the support set. Thereafter, a given fuzzy set will be expressed by its central value with an implicit support width, i.e. $A_i = \bar{e}_i^*$ and $B_j = \bar{c}_j^*$. Linguistically, the above expression may be interpreted as "about $\bar{e}_i^*(\bar{c}_j^*)$" in which "about" is constrained by the width value $\delta_i^e(\delta_j^c)$ of the corresponding support set.

3. *Expression of rules and data*

Suppose that there are $K$ rules in the rule-base denoted as

$$R_1 \ OR \ R_2 \ OR \ \cdots \ OR \ R_K$$

Each rule $R_k$ has the form of

$$R_k\text{: } IF \ \bar{e}_k^* \ AND \ \bar{c}_k^* \ THEN \ u_k \quad k \in [1, K] \tag{5.4}$$

where $\bar{e}_k^*$ is the central value corresponding to one of the predefined fuzzy sets $A_i$, $i \in [1, N_f]$, $\bar{c}_k^*$ is the central value in accordance with one of the predefined fuzzy sets $B_j$, $j \in [1, M_f]$, and $u_k$ is a real number in $\bar{U}$. Here no linguistic label is attached to the THEN part of the rule.

If we consider $\bar{e}_k^*$ and $\bar{c}_k^*$ as a 2-dimensional vector $r_k = (\bar{e}_k^*, \bar{c}_k^*)$, then $r_k$ will create a rectangle in the $\bar{E} \times \bar{C}$ plane taking $(\bar{e}_k^*, \bar{c}_k^*)$ as its central point and $2\delta_k^e$ and $2\delta_k^c$ as its sides. This is shown in Figure 5.1. Therefore, $K$ rules partition the

$\bar{E} \times \bar{C}$ plane into $K$ subplanes. It should be noted that the K subplanes are allowed to overlap because of the effect of fuzziness.

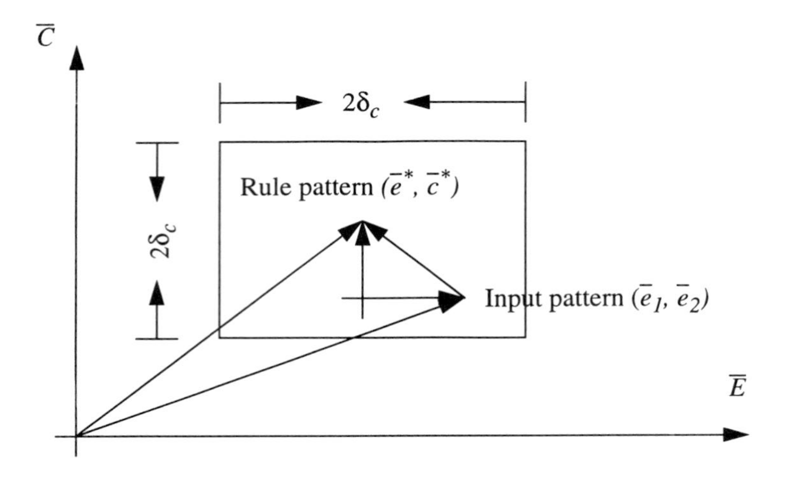


**Figure 5.1** Samples of the rule pattern and input pattern in input space

Similarly, current inputs $\bar{e}_I$ and $\bar{c}_I$ mapped from measured values $\bar{e}_m$ and $\bar{c}_m$ by scaling factors $GE$ and $GC$ can be expressed as a 2-dimensional vector $(\bar{e}_I, \bar{c}_I)$ in the $\bar{E}\times\bar{C}$ plane as shown in Figure 5.1.

4. *Pattern matching*

   The control algorithm can be viewed as a process in which an appropriate control action is deduced from a current input $(\bar{e}_I, \bar{c}_I)$ and $K$ rules according to some prespecified reasoning algorithms. In this approach, the reasoning procedure consists of two steps: pattern matching and weighted average. The first operation deals with the IF part for all rules, whereas the second one involves an operation on the THEN part for the fired rules.

   It is convenient to view the IF part of a rule and an input as patterns to be called a rule pattern and an input pattern respectively. In this view, the first step in the reasoning algorithm is to compute the matching degree between the input pattern and the rule patterns. Here we adopt the following Hamming distance algorithm.

   Let

$$l_k = (l_k^1, l_k^2) = (|\bar{e}_k^* - \bar{e}_I|\ , |\bar{c}_k^* - \bar{c}_I|) \tag{5.5}$$

be a minus vector. Then relative Hamming distance $d_k \in [0, 1]$ from input pattern $(\bar{e}_I, \bar{c}_I)$ to the $k$th rule pattern is given by

$$d_k = \frac{l_k^1 + l_k^2}{\delta_k^e + \delta_k^c} \tag{5.6}$$

where $l_k^1 \le \delta_k^e$ and $l_k^2 \le \delta_k^c$. The matching degree $s_k \in [0,1]$ is calculated by

$$s_k = 1 - d_k \tag{5.7}$$

It is evident from equations (5.6) and (5.7) that: if the input pattern is fully matched with the $k$th rule pattern, meaning that the input vector is exactly the same as the central value vector of the $k$th rule, then $d_k = 0$ and $s_k = 1$; if they are completely unmatched, implying that the input vector is on one of the corners of the rectangle created by the $k$th rule pattern, then $d_k = 1$ and $s_k = 0$; otherwise, $0 < s_k < 1$, indicating that there exists some similarity between the two patterns. Note that if the input pattern falls outside the $k$th rectangle, we define $s_k = 0$. Although a Hamming distance produces a diamond contour, the patterns falling into the effective area created by equation (5.6) is in fact a rectangle (Figure 5.1) inside the diamond determined by $\delta_k^e + \delta_k^c$. This results from the conditions of $l_k^1 \le \delta_k^e$ and $l_k^2 \le \delta_k^c$.

5. *Weighted average*
   Recall that the rule $R_k$ is expressed as "IF $(\bar{e}_k^*, \bar{c}_k^*)$ THEN $u_k$". If $s_k = 1$, the deduced control value should be $u = u_k$. In contrast, the rule $R_k$ has no contribution to the output if $s_k = 0$. However, if $0 < s_k < 1$, there are at least two rules determining the present control value.
   Suppose that for a specified input, after the matching process with $K$ rules, there exist $Q$ matching degrees satisfying $0 < s < 1$ and they are relabelled as $s_1, s_2, \cdots, s_Q$ with corresponding $Q$ THEN parts denoted as $u_1, u_2, \cdots, u_Q$. Then the present control value is given by

$$u = \frac{\sum_{q=1}^{Q} s_q \cdot u_q}{\sum_{q=1}^{Q} s_q} \tag{5.8}$$

Compared with the reasoning algorithms presented in Chapter 2, the algorithm described above is considerably simple and thus, computationally effective. Two basic procedures, fuzzification and defuzzification in normal fuzzy control algorithms, are not explicitly presented in the above algorithm.

## 5.3 Extracting rules from recorded data

### 5.3.1 Description of the methodology

As described in Chapter 4, there exist two kinds of error signals in the proposed system. One of them is the difference between the desired response and the process output. This error $e$, called the learning error, is used to learn correct control action u and will tend to zero in the time interval of interest with increase of iteration number. The other error $\bar{e}_m$, called the measured control error and defined as the difference between the set-point and the process output, is primarily employed to construct the rule-base which will be subsequently used in the fuzzy control system. It should be pointed out that $\bar{e}_m$ always undergoes a time course at each iteration consisting of a transient stage and a steady-state stage. The details of the approach are described next.

1. *Measurement of variables*

Suppose that, at the $k$th learning iteration, correct control actions $u_1$ $(iT_s)$ and $u_2$ $(iT_s)$ are learned with which the desired output responses specified by the reference models are achieved, where $T_s$ is sampling period and $i \in [0, L]$ with $L$ being the maximum sampling number. At the same time, two sets of inputs, $(\bar{e}_{m1}(iT_s), \bar{c}_{m1}(iT_s))$ and $(\bar{e}_{m2}(iT_s), \bar{c}_{m2}(iT_s))$, for two control loops are recorded. Here,

$$\begin{aligned} \bar{e}_{m1}(iT_s) &= SP_1 - y_1(iT_s) \\ \bar{c}_{m1}(iT_s) &= \bar{e}_{m1}(iT_s) - \bar{e}_{m1}(iT_s - T_s) \\ \bar{e}_{m2}(iT_s) &= SP_2 - y_2(iT_s) \\ \bar{c}_{m2}(iT_s) &= \bar{e}_{m2}(iT_s) - \bar{e}_{m2}(iT_s - T_s) \end{aligned} \tag{5.9}$$

for $i = 0, 1, 2, \cdots, L$, where $SP_1$ and $SP_2$ denote set-points for loop 1 and loop 2 respectively.

With corresponding $u_1(iT_s)$ and $u_2(iT_s)$, two sets of input–output pairs relating the input to the control action can be constructed in order as follows.

$$\begin{aligned} \left\{ \bar{e}_{m1}(iT_s)\ ,\ \bar{c}_{m1}(iT_s) \right\} &\sim \left\{ u_1(iT_s) \right\} \\ \left\{ \bar{e}_{m2}(iT_s)\ ,\ \bar{c}_{m2}(iT_s) \right\} &\sim \left\{ u_2(iT_s) \right\} \end{aligned} \tag{5.10}$$

for $i = 0, 1, 2, \cdots, L$, where $u_1(iT_s) \in [u_{min\,1}, u_{max\,1}]$ and $u_2(iT_s) \in [u_{min\,2}, u_{max\,2}]$.

2. *Scaling process*

Define two sets of scaling factors $(GE_1, GC_1)$ and $(GE_2, GC_2)$ so that the following is satisfied:

$$\begin{aligned} (\bar{e}_1(iT_s)\ ,\ \bar{c}_1(iT_s)) &\in \bar{E}_1 \times \bar{C}_1 \\ (\bar{e}_2(iT_s)\ ,\ \bar{c}_2(iT_s)) &\in \bar{E}_2 \times \bar{C}_2 \end{aligned} \tag{5.11}$$

where $\bar{E}$ and $\bar{C}$ are the universes of discourse defined by equation (5.1), and

$$\begin{aligned}
\bar{e}_1(iT_s) &= Q\,[\bar{e}_{m1}(iT_s)\cdot GE_1] \\
\bar{c}_1(iT_s) &= Q\,[\bar{c}_{m1}(iT_s)\cdot GC_1] \\
\bar{e}_2(iT_s) &= Q\,[\bar{e}_{m2}(iT_s)\cdot GE_2] \\
\bar{c}_2(iT_s) &= Q\,[\bar{c}_{m2}(iT_s)\cdot GC_2]
\end{aligned} \tag{5.12}$$

where $Q$ denotes an operation with which the scaled values are set to the nearest elements in $\bar{E}$ or $\bar{C}$. Since the scaling factor plays an important role in the fuzzy controller, we will discuss this issue in detail later.

Now, we have the following ordered pairs.

$$\begin{aligned}
\left\{ \bar{e}_1(iT_s),\, \bar{c}_1(iT_s) \right\} &\sim \left\{ u_1(iT_s) \right\} \\
\left\{ \bar{e}_2(iT_s),\, \bar{c}_2(iT_s) \right\} &\sim \left\{ u_2(iT_s) \right\}
\end{aligned} \tag{5.13}$$

for $i = 0, 1, \cdots, L$. In what follows, we will drop subscripts 1 and 2 and only deal with one set of data given in equation (5.13).

3. *Conflict resolution*

In order to create the rules similar to the form of equation (5.4), it is necessary to transform these $L + 1$ time-ordered data into a set of value-based input-output data in which there is no time index $iT_s$ involved.

First, $L + 1$ data pair are rearranged into J groups according to their vector values $(\bar{e}(iT_s),\bar{c}(iT_s))$ denoted as follows:

$\Gamma_j$:

$$\begin{aligned}
(\bar{e}\,,\,\bar{c})_{\Gamma_j} &\sim u_{\Gamma_j}^1 \\
(\bar{e}\,,\,\bar{c})_{\Gamma_j} &\sim u_{\Gamma_j}^2 \\
&\;\;\vdots \\
(\bar{e}\,,\,\bar{c})_{\Gamma_j} &\sim u_{\Gamma_j}^P
\end{aligned} \tag{5.14}$$

where $j = 1, 2, \cdots, J$. Notice that a single vector $(\bar{e}\,,\,\bar{c})_{\Gamma_j}$ may be associated with several different control values, $u_{\Gamma_j}^1, u_{\Gamma_j}^2, \cdots, u_{\Gamma_j}^P$. A data group $\Gamma_j$, $j = 1, 2, \cdots, J$, is said to be conflict if $P \neq 1$. We need to solve this conflict problem and it is referred to as conflict resolution.

A simple and reasonable method for solution is to take the average of $u_{\Gamma_j}^1, u_{\Gamma_j}^2, \cdots, u_{\Gamma_j}^P$ as their typical value provided that the properties possessed by these data are not taken into consideration, that is,

$$u_{\Gamma_j} = \frac{1}{P}\sum_{p=1}^{P} u_{\Gamma_j}^p \tag{5.15}$$

However, by observing and inspecting the error response subject to the step command signal we can conclude that the conflict is most likely to occur at both the beginning of the transient response stage and during the steady-state stage. Two reasons for this can be given. The pure time delay $h$ in control possessed by the controlled process is responsible for the former due to the fact that the error and change-in-error remain unchanged in the time interval $[0, hT_s]$ although the control action in $[0, hT_s]$ may take different values during the learning phase. In contrast, the discrete property of $\bar{E}$ and $\bar{C}$ and continuous characteristic of $\bar{U}$ cause the conflict during the steady state. It is evident that the average scheme is suitable for handling the latter situation because control action $u^p_{\Gamma_j}$ s during steady state can be viewed as some fluctuation above the average. However, the transient performance of the output will be degraded if this scheme is applied to the former case. An alternative method is to exclude $u(0), u(1), \cdots, u(h)$ from data set (5.10) and to use them directly as a set of initial control values when the fuzzy control algorithm is applied.

4. *Rule-base formation*
   Now, after the conflict resolution process, we obtain $J$ distinct data pairs

$$
\begin{aligned}
\Gamma_1 &: (\bar{e}_{\Gamma_1}, \bar{c}_{\Gamma_1}) \sim u_{\Gamma_1} \\
\Gamma_2 &: (\bar{e}_{\Gamma_2}, \bar{c}_{\Gamma_2}) \sim u_{\Gamma_2} \\
&\cdot \\
&\cdot \\
&\cdot \\
\Gamma_J &: (\bar{e}_{\Gamma_J}, \bar{c}_{\Gamma_J}) \sim u_{\Gamma_J}
\end{aligned}
\tag{5.16}
$$

Note that $u_{\Gamma_j}$, $j = 1, 2, \cdots, J$, need not to be distinct. If we assign some appropriate fuzzy sets predefined over $\bar{E}$ and $\bar{C}$ to $\bar{e}_{\Gamma_1}$ and $\bar{c}_{\Gamma_1}$, with $\bar{e}_{\Gamma_1}$ and $\bar{c}_{\Gamma_1}$ being corresponding central values, then the above data pairs can be thought of as a set of control rules with the understanding that the half width $\delta^e_j(\delta^c_j)$ is embedded implicitly.

Thus, we obtain $J$ control rules denoted as follows.

$$
\begin{aligned}
R_1 &: \; IF \; (\bar{e}^*_{\Gamma_1}, \bar{c}^*_{\Gamma_1}) \; THEN \; u_{\Gamma_1} \\
R_2 &: \; IF \; (\bar{e}^*_{\Gamma_2}, \bar{c}^*_{\Gamma_2}) \; THEN \; u_{\Gamma_2} \\
&\cdot \\
&\cdot \\
&\cdot \\
R_J &: \; IF \; (\bar{e}^*_{\Gamma_J}, \bar{c}^*_{\Gamma_J}) \; THEN \; u_{\Gamma_J}
\end{aligned}
\tag{5.17}
$$

where the symbol * is added to emphasize that $\bar{e}^*$ and $\bar{c}^*$ are fuzzy sets.

Notice that the present $J$ control rules are derived from either a positive or negative step command input. If it is assumed that the process being controlled has a skew symmetric input–output property about an operating point, that is, if the following equation is satisfied

$$f(\hat{u} + \Delta u) - f(\hat{u}) = f(\hat{u}) - f(\hat{u} - \Delta u) \tag{5.18}$$

where $f(\cdot)$ denotes the process input–output function, $f(\hat{u})$ is the operating point and $\Delta u \in [u_{\min}, u_{\max}]$ is an increment around the $\hat{u}$, then it is expected that, in accordance with an opposite sign step input, a set of data pairs having the same absolute values as in equation (5.13) but with opposite signs would be obtained. Taking into account the symmetric property of $\overline{E}$ and $\overline{C}$ and the assumption about the process made above, we derive another set of rules given by

$$\tilde{R}_j: \quad IF \quad (-\,\overline{e}^{*}_{\Gamma_j} \;, \quad -\overline{c}^{*}_{\Gamma_{Jj}} \;) \quad THEN \quad -u_{\Gamma_j} \tag{5.19}$$

for $j = 1, 2, \cdots, J$.

Therefore, corresponding to the positive and negative step commands, two sets of rules are built into the rule-base with numbers of rules being $2J$. It should be pointed out that it is possible to use only $J$ control rules instead of doubling the rule numbers. In this case, to ensure the completeness of the rule-base as discussed next, the control inputs not falling into the rule plane determined by the the learned $J$ rules must be negatively scaled with unity gains and the corresponding controller output must be scaled as well in the same manner.

5. *Rule-base verification*
The rule-base constructed following the procedures presented previously must be verified to make sure that it is trustworthy. There are a number of specifications employed to evaluate its quality. Some of them are discussed here.
*Completeness*: Completeness means that there always exists at least one rule which will be fired for any possible input data, implying that rules should be well distributed on the rule plane $\overline{E}\times\overline{C}$. In particular, given any input pattern $(\overline{e}_I, \overline{c}_I) \in \overline{E} \times \overline{C}$, there must exist at least one rule pattern, say $(\overline{e}^{*}_k, \overline{c}^{*}_k) \in \overline{E} \times \overline{C}$, in the rule-base with which the matching degree between those two patterns is more than zero, i.e. $0 < s_k \leq 1$. This requirement is of potential importance because non-completeness of the rule-base may result in a situation in which no control action is taken during some control cycles. However, it should be emphasized that the completeness for a fuzzy rule-base does not mean that there must exist a rule for all possible combinations of fuzzy sets defined over $\overline{E}$ and $\overline{C}$. In contrast, completeness can be assured if the union of all subplanes created by the individual rule pattern covers the area in which all input patterns may occur. By choosing the appropriate width of the support sets and noting the problem-oriented property of the learning system, the completeness requirement can be satisfied in the sense discussed above.
*Correctness*: This requirement mainly involves the THEN part of the rule. It is guaranteed by the fact that the control action (THEN part) is derived by learning and is considered to be correct.

*Consistency*: The conflict resolution procedure discussed earlier ensures that inconsistent data pairs are excluded from the constructed rule-base. Unlike non-fuzzy reasoning systems, in our case it is allowed that one input pattern should fire several rules at the same time and this situation is not regarded as being inconsistent.

*Complexity*: Here complexity simply refers to the number of rules in the rule-base. It depends primarily upon the definition of the universes of discourse $\bar{E}$ and $\bar{C}$ (equation (5.1)), more specifically, upon the cardinalities of $\bar{E}$ and $\bar{C}$, i.e. $2N+1$ and $2M+1$ respectively. The larger $N(M)$ is, the larger the rule-base size and hence more complicated the rule-base. In addition, the rule number is affected by the transient performance specified during the learning stage. The learning mechanism with different requirements will produce different data sets which are the basis for building the rule-base. For example, more rules will be created if the overshoot index is larger because in this case more rule patterns have to be excited in the rule plane.

*Reproducibility*: Recall that, in the learning stage, the system performance indices specified by the reference models are the objectives for the learning system to achieve. The term reproducibility is used to indicate the capacity of the learned rule-based fuzzy controller to reproduce these performance indices. This ability depends not only on the rule-base, but also on the fuzzy control algorithm employed. In Section 5.4, this will be evaluated by some defined measures and illustrated by numerical results.

### 5.3.2 Selection of scaling factors

It is well recognized that the scaling factors $GE$, $GC$ and $GU$ play important roles in determining the performance of a fuzzy control system and there is no universal method to determine their values. However, the iterative operation in the learning system makes this determination easier because some parameters required for calculating the scaling factors can be obtained during the learning process. In what follows, two cases about this issue will be discussed, namely, the set-point in the fuzzy control stage is the same as, and then different from, that in the learning stage.

1. *The same case*
   As mentioned previously, the function of the input scaling factors $GE$ and $GC$ is to map the measured variables $\bar{e}_m$ and $\bar{c}_m$ into $\bar{e} \in \bar{E}$ and $\bar{c} \in \bar{C}$ respectively. In order to utilize the whole space $\bar{E} \times \bar{C}$, it is desirable to map the maximum absolute value of $\bar{e}_m(\bar{c}_m)$ into the maximum element in $\bar{E}$ $(\bar{C})$. By noting the property of linear mapping, we have

$$GE = \frac{v_N}{|\bar{e}_m|_{\max}}$$
$$GC = \frac{w_M}{|\bar{c}_m|_{\max}} \tag{5.20}$$

where $v_N > 0$ and $w_M > 0$ are the maximum elements in $\bar{E}$ and $\bar{C}$ respectively, $|\bar{e}_m|_{\max}$ and $|\bar{c}_m|_{\max}$ are the maximum measured error and change-in-error respectively. Notice that if, without loss of generality, we set the initial value of the process output $y$ to be zero, then we have $|\bar{e}_m|_{\max} = |SP|$ and therefore, $GE$ can be determined. However, in general, $|\bar{c}_m|_{\max}$ cannot easily be decided in advance. Fortunately, it can be obtained during the learning phase. At the end of each iteration, all the data of $\bar{c}_m$ are available and the maximum absolute value denoted as $\bar{c}_L$ is found and is used as $|\bar{c}_m|_{\max}$.

With regard to the output scaling factor $GU$, we simply set it to 1 because in the process of constructing the rule-base no scaling operation is imposed on the learned control action $u$.

To summarize, in this simple case, we obtain

$$GE = \frac{v_N}{|SP|}$$
$$GC = \frac{w_M}{\bar{c}_L} \tag{5.21}$$
$$GU = 1$$

2. *The different case*

Equation (5.21) is applicable only if the set-points are the same in both the learning and control stages. However, in some cases, we wish to change the set-point and do not want to repeat the learning process. This objective may be achieved by modifying the scaling factors only, with the rule-base built from previous learning remaining unchanged.

Let $GE$, $GC$ and $GU$ be determined by equation (5.21) corresponding to $SP$. By the maximum mapping principle, the modified input scaling factors $\tilde{GE}$ and $\tilde{GC}$ corresponding to the changed set-point $\tilde{SP}$ are given by

$$\tilde{GE} = 1/K_{sp} \cdot GE$$
$$\tilde{GC} = K_{sp} \cdot GC \tag{5.22}$$

where $K_{sp} = |\, \tilde{SP}/SP \,|$.

Because of interaction effects between the two control loops, the determination of the output factor $\tilde{GU}$ is slightly difficult and requires some knowledge about the controlled process. Here it is assumed that the steady-state gain matrix is available either from *a priori* knowledge or through on-line estimation during the learning stage, as will be described later.

Let the steady-state gain matrix $\Lambda$ relating the two process inputs to the two outputs be given by

$$\Lambda = \begin{bmatrix} \lambda_{11} & \lambda_{12} \\ \lambda_{21} & \lambda_{22} \end{bmatrix} \tag{5.23}$$

and its inverse matrix $\Psi$ be given by

$$\Phi = \Lambda^{-1} = \begin{bmatrix} \psi_{11} & \psi_{12} \\ \psi_{21} & \psi_{22} \end{bmatrix} \tag{5.24}$$

It can be proved easily that, providing only the steady-state is considered, the modified $\tilde{GU}_1$ and $\tilde{GU}_2$ are given by

$$\begin{aligned} \tilde{GU}_1 &= K_{sp\,1}\cdot(\frac{1+\gamma_1\alpha}{1+\gamma_1\beta})\cdot GU_1 = K_{sp\,1}\cdot(\frac{1+\gamma_1\alpha}{1+\gamma_1\beta}) \\ \tilde{GU}_2 &= K_{sp\,2}\cdot(\frac{1+\frac{\gamma_2}{\alpha}}{1+\frac{\gamma_2}{\beta}})\cdot GU_2 = K_{sp\,2}\cdot(\frac{1+\frac{\gamma_2}{\alpha}}{1+\frac{\gamma_2}{\beta}}) \end{aligned} \tag{5.25}$$

where

$$\begin{aligned} K_{sp\,1} &= \frac{\tilde{SP}_1}{SP_1} \qquad & K_{sp\,2} &= \frac{\tilde{SP}_2}{SP_2} \\ \alpha &= \frac{\tilde{SP}_2}{\tilde{SP}_1} \qquad & \beta &= \frac{SP_2}{SP_1} \\ \gamma_1 &= \frac{\psi_{12}}{\psi_{11}} \qquad & \gamma_2 &= \frac{\psi_{21}}{\psi_{22}} \end{aligned} \tag{5.26}$$

Here subscripts 1 and 2 refer to control loop 1 and loop 2 respectively.

The gain matrix $\Lambda$ can be estimated while the learning is progressing. Suppose that the steady-states of $y_1$, $y_2$, $u_1$ and $u_2$ at the $k$th learning iteration are denoted as $y_{1s}^k$, $y_{2s}^k$, $u_{1s}^k$ and $u_{2s}^k$ and they satisfy the following:

$$\begin{bmatrix} y_{1s}^k \\ y_{2s}^k \end{bmatrix} = \begin{bmatrix} \lambda_{11} & \lambda_{12} \\ \lambda_{21} & \lambda_{22} \end{bmatrix} \begin{bmatrix} u_{1s}^k \\ u_{2s}^k \end{bmatrix} \tag{5.27}$$

where the gain matrix $\Lambda = \{\lambda_{ij}\}$ is assumed to be unknown. Likewise, at the $l$th iteration, l$\neq k$, we have

$$\begin{bmatrix} y_{1s}^l \\ y_{2s}^l \end{bmatrix} = \begin{bmatrix} \lambda_{11} & \lambda_{12} \\ \lambda_{21} & \lambda_{22} \end{bmatrix} \begin{bmatrix} u_{1s}^l \\ u_{2s}^k \end{bmatrix} \tag{5.28}$$

Let $\lambda = [\lambda_{11}, \lambda_{12}, \lambda_{21}, \lambda_{22}]^T$, $y_s = [y^k_{1s}, y^k_{2s}, y^l_{1s}, y^l_{2s}]^T$ and

$$u_s = \begin{bmatrix} u^k_{1s} & u^k_{2s} & 0 & 0 \\ 0 & 0 & u^k_{1s} & u^k_{2s} \\ u^l_{1s} & u^l_{2s} & 0 & 0 \\ 0 & 0 & u^l_{1s} & u^l_{2s} \end{bmatrix} \tag{5.29}$$

We have

$$y_s = u_s \cdot \lambda \tag{5.30}$$

It can be verified that $u_s$ is a full rank matrix, i.e. rank$[u_s] = 4$, if

$$u^k_{1s} \cdot u^l_{2s} - u^l_{1s} \cdot u^k_{2s} \neq 0 \tag{5.31}$$

Hence equation (5.30) has a unique solution, if the condition (5.31) is satisfied, given by

$$\lambda = u_s^{-1} \cdot y_s \tag{5.32}$$

which can be used to calculate $\tilde{GU}_1$ and $\tilde{GU}_2$ via equation (5.25).

## 5.4 Application to multivariable blood pressure control

The problem of simultaneous regulation of blood pressure (MAP) and cardiac output (CO) has been studied using the proposed method. Before presenting the simulation results using the model given in Appendix II, computing procedures are given and the performance measures are defined in the next two subsections.

### 5.4.1 Computing procedures

Basically, the computational process consists of three main steps which are summarized as follows.

Step 1: Apply the learning algorithm.

Step 2: Construct rule-bases.

Step 3: Determine whether to stop learning or not. If no, go to Step 1; otherwise, continue.

Step 4: Apply the fuzzy control algorithm.

More details of the above steps are described next. In what follows, we assume that *CO*/DOP and *MAP*/SNP are paired, comprising two control loops. This pairing is consistent with the results derived from conceptual considerations in which DOP is primarily used to increase $CO$, and SNP is mainly aimed at lowering $MAP$.

1. *Learning algorithm*

   Three learning algorithms similar to those presented in Chapter 4 are adopted in the simulations.

   (a) Error correction:

$$
\begin{aligned}
u_{co}^{k+1}(i) &= u_{co}^{k}(i) + g_1 e_{co}^{k}(i+h) \\
u_{map}^{k+1}(i) &= u_{map}^{k}(i) + g_2 e_{map}^{k}(i+h)
\end{aligned}
\tag{5.33}
$$

   where

$$
\begin{aligned}
e_{co}^{k}(i+h) &= y_{co}^{d}(i+h) - y_{co}^{k}(i+h) \\
e_{map}^{k}(i+h) &= y_{map}^{d}(i+h) - y_{map}^{k}(i+h)
\end{aligned}
\tag{5.34}
$$

   $k$ denotes iteration number, $h$ is an estimated delay in control, $g_1$ and $g_2$ are learning gains and superscript $d$ means the desired value.

   (b) Error and change-in-error correction:

$$
\begin{aligned}
u_{co}^{k+1}(i) &= u_{co}^{k}(i) + Cg_1 \cdot g_1 \cdot e_{co}^{k}(i+h) + (1-Cg_1) \cdot g_1 \cdot c_{co}^{k}(i+h) \\
u_{map}^{k+1}(i) &= u_{map}^{k}(i) + Cg_2 \cdot g_2 \cdot e_{co}^{k}(i+h) + (1-Cg_2) \cdot g_2 \cdot c_{map}^{k}(i+h)
\end{aligned}
\tag{5.35}
$$

   where

$$
\begin{aligned}
c_{co}^{k}(i+h) &= e_{co}^{k}(i+h) - e_{co}^{k}(i+h-1) \\
c_{map}^{k}(i+h) &= e_{map}^{k}(i+h) - e_{map}^{k}(i+h-1)
\end{aligned}
\tag{5.36}
$$

   are the changes of learning errors and $0 \le Cg_1 \le 1$, $0 \le Cg_2 \le 1$ are used to assign the proportions of error and change-in-error contributing to the learning algorithm.

   (c) Cross-error correction:

$$
\begin{aligned}
u_{co}^{k+1}(i) &= u_{co}^{k}(i) + Cg_1 \cdot g_1 \cdot e_{co}^{k}(i+h) + (1-Cg_1) \cdot g_1 \cdot e_{map}^{k}(i+h) \\
u_{map}^{k+1}(i) &= u_{map}^{k}(i) + Cg_2 \cdot g_2 \cdot e_{map}^{k}(i+h) + (1-Cg_2) \cdot g_2 \cdot e_{co}^{k}(i+h)
\end{aligned}
\tag{5.37}
$$

   where again $0 \le Cg_1 \le 1$, $0 \le Cg_2 \le 1$ are used to assign the proportions of error and cross-error contributed to the learning algorithm.

2. *Desired response*

   The desired responses $y_{co}^{d}$ and $y_{map}^{d}$ are obtained from the outputs of the reference models subject to step command signals and are given, in the $s$-domain, by equations (5.38) and (5.39) below:

(a) First-order model response:

$$Y_{co}(s) = \frac{e^{-\tau_{co} s}}{T_{co} s + 1} \cdot \frac{SP_{co}}{s}$$
$$Y_{map}(s) = \frac{e^{-\tau_{map} s}}{T_{map} s + 1} \cdot \frac{SP_{map}}{s} \tag{5.38}$$

where $SP$ denotes the set-point, $T$ represents the time constant and $\tau$ is the time delay in control. Note that equation (5.38) implies that the steady-state error is zero and the transient specification can be determined by the value of the time constant.

(b) Second-order model response:

$$Y_{co}(s) = \frac{\omega_{nco}^2 \cdot e^{-\tau_{co} s}}{s^2 + 2\xi_{co}\,\omega_{nco} + \omega_{nco}^2} \cdot \frac{SP_{co}}{s}$$
$$Y_{map}(s) = \frac{\omega_{nmap}^2 \cdot e^{-\tau_{map} s}}{s^2 + 2\xi_{map}\,\omega_{nmap} + \omega_{nmap}^2} \cdot \frac{SP_{map}}{s} \tag{5.39}$$

where $\xi$ and $\omega_n$ are the damping factor and the natural frequency respectively. The desired transient performance can be determined by the values of $\xi$, which is closely associated with overshoot, and $\omega_n$ or equivalently $t_s$ (settling time) through an approximate relationship $t_s = 4/\ \xi w_n$. In other words, the pole positions in the s-plane are assigned by the values of $\xi$ and $\omega_n$.

3. *Definition of fuzzy sets*
For the sake of simplicity, all the universes of discourse used in both rule-base construction and the fuzzy control stage are defined to be the same, that is, $\bar{E}_{co} = \bar{E}_{map} = \bar{C}_{co} = \bar{C}_{co} =$

$$\left\{-6,-5,-4,-3,-2,-1,0,1,2,3,4,5,6\right\}$$

Again, for simplicity, thirteen fuzzy sets are defined over $\bar{E}$ and $\bar{C}$, which may represent linguistic labels ranging from *negative_very_large* through *zero* to *positive_very_large* with the central values ranging from −6 to +6. The widths of all the fuzzy sets are taken to be the same, i.e. $\delta_e = \delta_c = 3$.

4. *Criteria for stopping learning*
Recall that, as presented in Chapter 4, an infinite number of iterations are required in order to make the learning error be zero in the time interval of interest $[0, L]$. However, in practice, a finite iteration number is expected and some criteria must be adopted to stop the learning process. We propose the following three possible criteria. More specifically, the learning process will be terminated:

(a) When one of the performance measures (defined later on) reaches a lower value near 0 or higher value near 1 and remains relatively stable; or

(b) When $L$-2 norm $\| e \|_2$ or infinity norm $\| e \|_\infty$ is less than a small prespecified value; or

(c) When the number of rules in the constructed rule base remains unchanged or little changed.

### 5.4.2 Performance assessment

The results of the learning control and the fuzzy control can be evaluated in a number of ways. Besides directly inspecting the output response curves, in this chapter some numerical measures are defined used to assess the performance.

First, we define the following errors.

$$\begin{aligned} \varepsilon_d &= SP - y_d \\ \varepsilon_f &= SP - y_f \\ \varepsilon_l^k &= SP - y_l^k \end{aligned} \tag{5.40}$$

where $y_d$, $y_f$ and $y_l^k$ denote the desired, the fuzzy controlled and the $k$th learned process output responses with $y$ being referred to either $y_{co}$ or $y_{map}$. Notice that, by the definition, $\varepsilon_f = \bar{e}_m$, the measured control error, and $\varepsilon_l^k - \varepsilon_d = e$, the learning error.

Second, two error criteria *ISE* and *ITAE* as defined in equation (3.16) will be used. The instant error appearing in the integrals can be one of the errors defined in equation (5.40). We note that *ISE* and *ITAE* have different physical interpretations. The *ISE* weights large errors usually occurring at the beginning of the transient stage, whereas the *ITAE* weights heavily errors of longer duration which occur frequently during the steady-state stage. Thus, we may consider the *ISE* as an appropriate indicator for the transient performance and the *ITAE* for the steady-state performance.

Finally, denoting

$$\Phi_d = \left\{ ISE\_CO_d,\ ITAE\_CO_d,\ ISE\_MAP_d,\ ITAE\_MAP_d \right\}$$

$$\Phi_f = \left\{ ISE\_CO_f,\ ITAE\_CO_f,\ ISE\_MAP_f,\ ITAE\_MAP_f \right\}$$

$$\Phi_l^k = \left\{ ISE\_CO_l^k,\ ITAE\_CO_l^k,\ ISE\_MAP_l^k,\ ITAE\_MAP_l^k \right\}$$

the following performance measures are defined:

1. Absolute and relative differences between the learned and the desired performances (*AD* and *RD*)

$$AD_i^k = |\phi_{li}^k - \phi_{di}^k|$$
$$RD_i^k = \frac{|\phi_{li}^k - \phi_{di}^k|}{\phi_{di}^k} \tag{5.41}$$

   where $i = 1, 2, 3, 4$, $\phi_{li}^k \in \Phi_l^k$, and $\phi_{di} \in \Phi_d$. *AD* and *RD* are primarily used to compare the learning speed in terms of the performance errors when the different learning gains, initial controls and learning algorithms are performed.

2. Performance improvement factor (*IP*)

$$IP_i^k = 1 - \frac{|\phi_{di} - \phi_{li}^k|}{|\phi_{di} - \phi_{li}^1|} \tag{5.42}$$

   where $i = 1, 2, 3, 4$, $\phi_{li}^k \in \Phi_l^k$, $\phi_{di} \in \Phi_d$, and $\phi_{li}^1 \in \Phi_l^1$ is the index produced at the first iteration. *IP* is employed to measure the learning ability with respect to a variety of situations such as the change of process parameters and the change of the desired responses.

3. The relative difference between the rule-based fuzzy controlled and the learned responses (*FRD*).

$$FRD_i = \frac{|\phi_{li}^k - \phi_{fi}|}{\phi_{li}^k} \tag{5.43}$$

   where $i = 1, 2, 3, 4$, $\phi_{fi} \in \Phi_f$ and $\phi_{li}^k \in \Phi_l^k$ with $k$ being the iteration number at which the rule-base is constructed. *FRD* is adopted to assess the reproducibility of the fuzzy controller in which the rule-base is built from the $k$th learned control actions.

### 5.4.3 Simulation results

The situations concerned in this simulation study are divided into three main groups: learning ability/reproducibility, the comparisons of convergence speed/reproducibility and learning robustness. The first group of simulations is aimed at testing the adaptability of the learning algorithm with respect to the variation of the desired response and the process parameters, whereas the intention of the second group is to investigate how the convergency rate of the learning system is affected by the learning gains, initial control action and different learning update laws. Along with all simulation cases, reproducibility of the resultant fuzzy control system, which is schematically shown in Figure 5.2, in terms of the control performances is evaluated, for which the simplified control algorithm presented previously is used. The purpose of the last group of simulations is to examine how robust the learning algorithm is with respect to the estimated time delay and to a noisy environment.

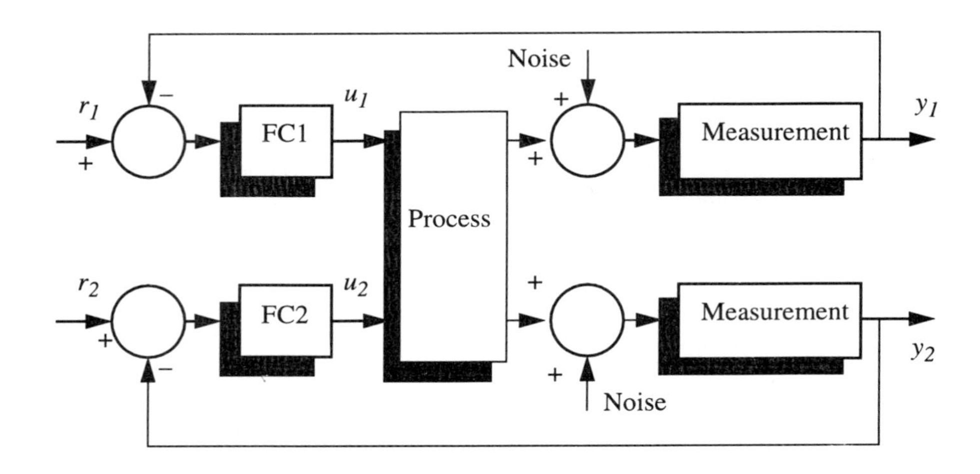


**Figure 5.2** Resultant multivariable fuzzy control system

The following values are used throughout the simulations.

Sampling period $T_s$ = 30 s
Learning interval $L$ = 100
Maximum iteration number $K$ = 15
Initial condition $y_{co}^k(0) = y_{map}^k(0) = 0$
Normal value of $co$: $co_0$ = 100 ml/s
Normal value of $map$: $map_0$ = 120 mmHg
Set-point for $co$: $SP_{co}$=20 ml/s
Set-point for $map$: $SP_{map}$ = −10 mmHg
Time delay in the reference model: $\tau_{co} = \tau_{map} = 3T_s$;

In addition, it is assumed that the rule-base is constructed based on the data obtained at the tenth iteration.

1. *Learning ability/reproducibility*
   Throughout this group of simulations, the following is assumed: only error-correction learning law (equation (5.33)) is used; initial control actions $u_{co}^0$, $u_{map}^0$ are set to be zero; the estimated time delay number $h$ in the learning law is chosen to be 5; and the learning gains are set to be $g_1$= 0.06 and $g_2$= −0.07.

(a) The first-order specifications
When the desired responses are specified by the first-order reference models, three cases are considered:

$$T_{co} = 60 \text{ s} \qquad T_{map} = 60 \text{ s}$$
$$T_{co} = 60 \text{ s} \qquad T_{map} = 120 \text{ s}$$
$$T_{co} = 120 \text{ s} \qquad T_{map} = 60 \text{ s}$$

where $T_{co}$ and $T_{map}$ denote time constants in the models (equation (5.38)) for *CO* and *MAP* respectively.

By taking the parameters mentioned above, three sets of simulations were performed. Figure 5.3 shows one of the results, where $T_{co}$ = 120 s and $T_{map}$ = 60 s. The desired, learned (at the first, the fifth and the tenth iterations) and fuzzy controlled $y_{co}$ and $y_{map}$ are displayed in the Figure. It can be seen that the learned response asymptotically approached the desired one with an increase of the iteration number. Interaction effects within the two control loops are well handled by the learned individual control actions $u_{dop}$ and $u_{snp}$ which are also illustrated in Figure 5.3.

Learning ability is also demonstrated in Figure 5.4 by the fact that the improvement factors (here only the results corresponding to *MAP* are given) under different desired specifications approach 1 after a few iterations. It can be seen from the figures that the transient responses (indicated by *ISE*) improved more quickly than the steady state responses (indicated by *ITAE*). However, after a few iterations not much improvement in the transient responses took place, while the steady-state responses were improved further. Reproducibility of the resulted fuzzy controller can be verified by comparing the learned response at the tenth iteration with the fuzzy controlled response in Figure 5.3. The relative difference $FRD_i$ defined in the previous subsection corresponding to Figure 5.3 are presented in Table 5.1 where small values imply that a higher accuracy is possessed by the constructed rule-bases with respect to the learned control actions. The number of rules for loop 1 and loop 2 with the iteration number are shown in Table 5.2. It can be seen that the number of rules remained unchanged after seven iterations with the numbers being 18 and 10 respectively.

(b) The second-order specifications
There are two parameters to be specified in the second-order reference models. We consider three situations:

$$\xi_{co} = 0.8 \quad t_{s,co} = 240 \text{ s} \quad \xi_{map} = 0.8 \quad t_{s,map} = 240 \text{ s}$$
$$\xi_{co} = 0.9 \quad t_{s,co} = 240 \text{ s} \quad \xi_{map} = 0.7 \quad t_{s,map} = 300 \text{ s}$$
$$\xi_{co} = 0.7 \quad t_{s,co} = 300 \text{ s} \quad \xi_{map} = 0.9 \quad t_{s,map} = 240 \text{ s}$$

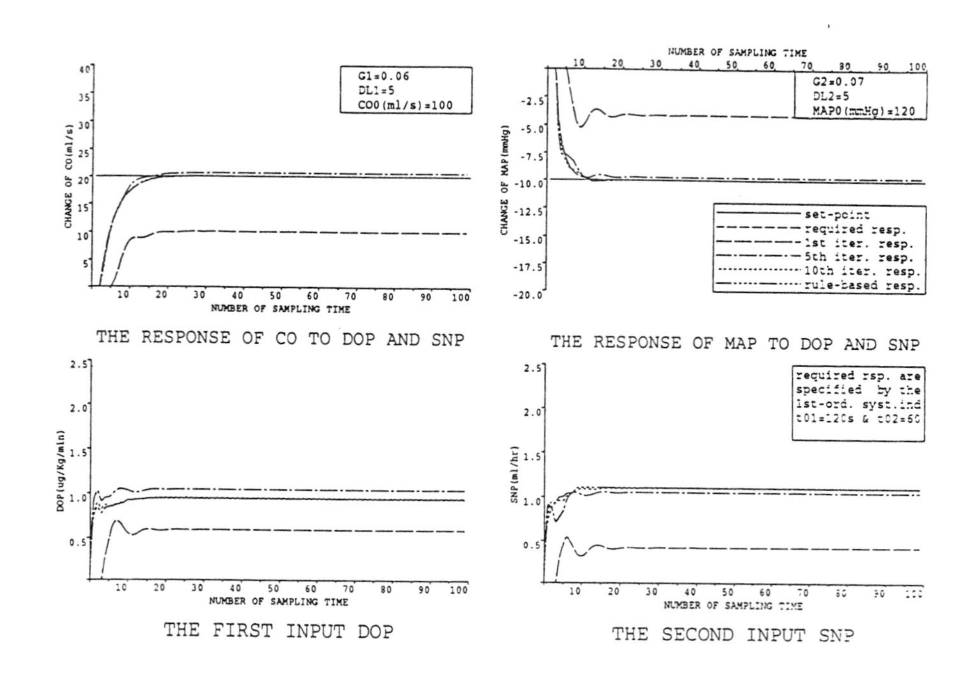


**Figure 5.3** Responses of the learning and fuzzy control: first-order specifications

**Table 5.1** Reproducibility of fuzzy controller measured by *FRD*

| Fig. no. | $FRD_1 \times 100$ | $FRD_2 \times 100$ | $FRD_3 \times 100$ | $FRD_4 \times 100$ |
|---|---|---|---|---|
| 5.3 | 0.79 | 0.98 | 0.86 | 5.6 |
| 5.5 | 0.00 | 1.73 | 0.00 | 0.44 |
| 5.7 | 0.50 | 1.75 | 0.00 | 0.11 |
| * | 0.05 | 5.17 | 0.16 | 1.07 |
| # | 0.00 | 1.74 | 0.25 | 0.47 |

*: change of $k_{11}$
#: change of $T_2$

where $t_s$ denotes the settling time. Unlike the first-order model, more flexible transient responses can be experienced by selecting the values of $\xi$ and $t_s(\omega_n)$.

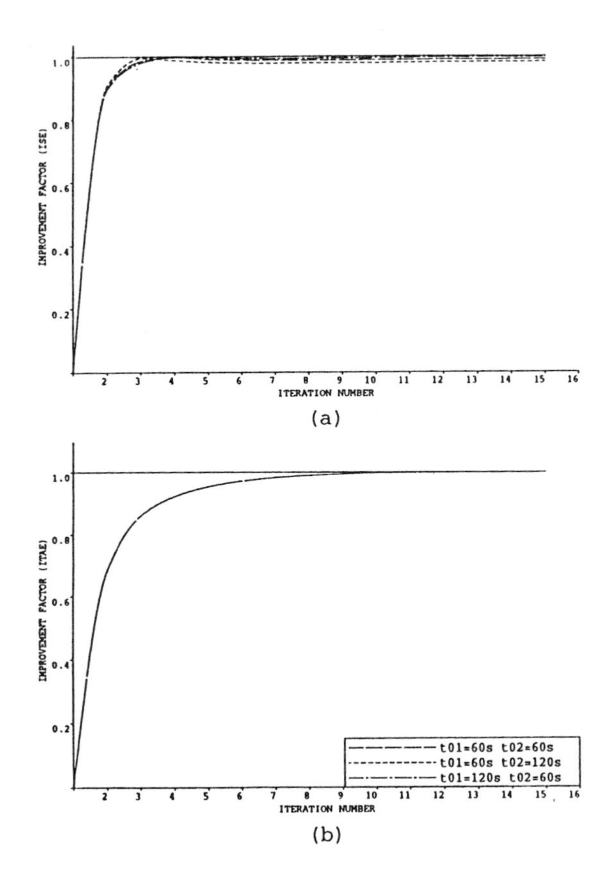


**Figure 5.4** Improvement factor *IP* for *MAP* during learning: first-order specifications

Small $\xi$ will result in a faster rise time and a larger overshoot. A good selection of $\xi$ is about 0.7.

Simulation results corresponding to the first case are depicted in Figure 5.5. As discussed in the previous subsection, both learning ability and reproducibility are illustrated in the figure. An important conclusion is that the desired transient response speed concerning the two loops can be made separately and achieved by the learning process despite the interaction effects. Improvement factors corresponding to *co* are shown in Figure 5.6 and the same comments as in the previous subsection can be made. *FRD* and the rule numbers corresponding to Figure 5.5 are presented in Table 5.1 and Table 5.2 respectively.

**Table 5.2** Rule numbers in rule-bases 1 and 2

| Loop number | Iteration number | | | | | | | | | | Correspond to Fig. no. |
|---|---|---|---|---|---|---|---|---|---|---|---|
| | 1 | 2 | 3 | 4 | 5 | 6 | 7 | 8 | 9 | 10 | |
| 1 | 12 | 20 | 20 | 18 | 16 | 18 | 18 | 18 | 18 | 18 | |
| 2 | 16 | 20 | 18 | 12 | 12 | 12 | 8 | 10 | 10 | 10 | 5.3 |
| Sum | 28 | 40 | 38 | 30 | 28 | 30 | 26 | 28 | 28 | 28 | |
| 1 | 14 | 20 | 24 | 24 | 18 | 16 | 16 | 16 | 16 | 16 | |
| 2 | 20 | 26 | 18 | 12 | 12 | 12 | 10 | 12 | 12 | 14 | 5.5 |
| Sum | 34 | 46 | 42 | 36 | 30 | 28 | 26 | 28 | 28 | 30 | |
| 1 | 14 | 22 | 24 | 20 | 16 | 14 | 16 | 16 | 16 | 16 | |
| 2 | 24 | 32 | 14 | 12 | 12 | 12 | 12 | 12 | 12 | 14 | 5.7 |
| Sum | 38 | 54 | 38 | 32 | 28 | 26 | 28 | 28 | 28 | 30 | |
| 1 | 14 | 22 | 22 | 18 | 16 | 16 | 16 | 16 | 16 | 16 | |
| 2 | 18 | 30 | 14 | 10 | 10 | 10 | 12 | 12 | 12 | 12 | * |
| Sum | 32 | 52 | 36 | 28 | 26 | 26 | 28 | 28 | 28 | 28 | |
| 1 | 16 | 20 | 24 | 24 | 20 | 16 | 16 | 16 | 14 | 14 | |
| 2 | 18 | 22 | 16 | 14 | 14 | 12 | 12 | 12 | 12 | 12 | # |
| Sum | 34 | 42 | 40 | 38 | 34 | 28 | 28 | 28 | 26 | 26 | |

*: change of $k_{11}$
#: change of $T_2$

(c) Variation of process parameters

In order to investigate the adaptability of the learning system with respect to the different process parameters, process gains and time constants were changed from their typical values by up to 20% increase or decrease. Here again, the learning gains were the same as before, i.e. $g_1$ = 0.06 and $g_2$ = –0.07. In addition, the desired responses were specified by the second-order models with $\xi_{co} = \xi_{map} = 0.8$ and $t_{s,co} = t_{s,map} = 240$ s. Although simulations for all of the gains and the time constants were performed, we show only one of them for illustration. Figure 5.7 shows the process outputs and inputs when gain $K_{22}$ increased 20% from its typical value. It appears that the learning system is capable of tracking the desired responses in spite of large variations of the process parameters. By comparing the results, we find that all the output responses with different process parameters are similar to each other but the learned control actions are substantially different, implying that the learning scheme is very efficient at dealing with different processes. This efficiency is also demonstrated by improvement factors (for *MAP*) as shown in Figure 5.8, where process gains were changed by up to 20%. The results of the fuzzy control indicated in Figure 5.7 once again demonstrate that the rule-based controller established from the learning system is capable of producing a

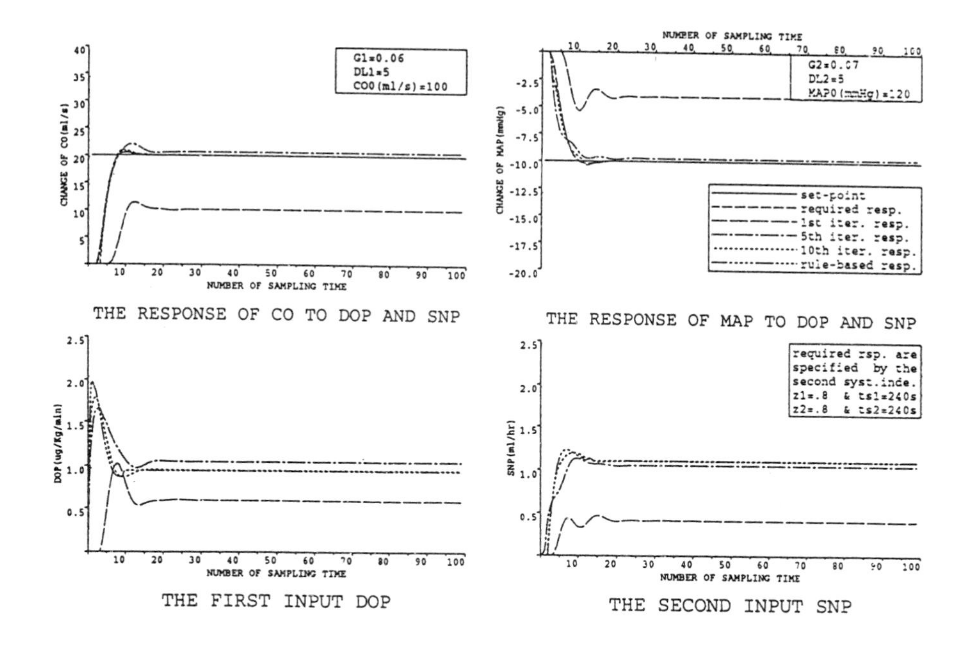


**Figure 5.5** Responses for second-order specifications: $\zeta_1 = \zeta_2 = 0.8$; $t_{s1} = t_{ts2} = 240$

performance as good as the learning system. *FRD* measures and rule numbers created corresponding to Figure 5.7 (the change of $k_{22}$), to the change of $k_{11}$, and to the change of $T_2$ are presented in Table 5.1 and Table 5.2 respectively.

2. *Learning speed/reproducibility*
   As indicated in Chapter 4, no specific theoretical results concerning the learning speed are given. This group of studies attempts to explore this issue by means of numerical simulations. Here the learning speed is defined as the required iteration number at which the desired performance measured by *AD* and *RD* defined in equation (5.41) is achieved. In general, there are three factors affecting the learning speed, namely, learning gains, initial control actions and learning update laws. In what follows, no detailed comments on reproducibility will be made because a number of simulations have suggested that this ability can almost always be ensured.

   (a) *Learning gains*
   The learning gains are perhaps the main factors in determining the learning speed. Assume that the desired responses are specified by the second-order

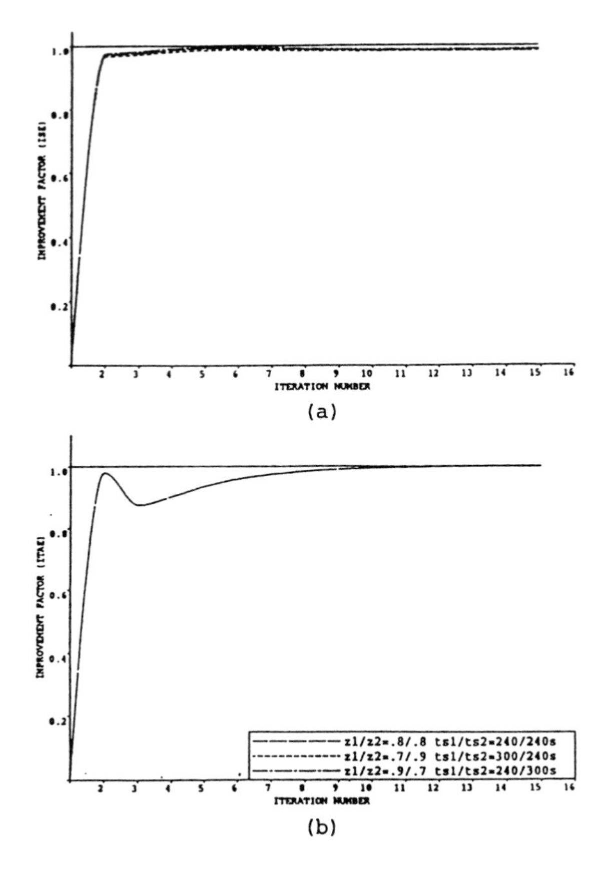


**Figure 5.6** Improvement factor *IP* for *CO* during learning: second-order specifications

models with $\xi_{co} = \xi_{map} = 0.8$ and $t_{s,co} = t_{s,map} = 240$ s and the error-correction update law is used. The following three gain sets were chosen for the simulations:

$$g_1 = 0.06 \qquad g_2 = -0.07$$
$$g_1 = 0.04 \qquad g_2 = -0.05$$
$$g_1 = 0.08 \qquad g_2 = -0.09$$

For the purpose of a clear comparison, here only a detailed version of the performance measures *AD* against the iteration number starting from 3 is shown

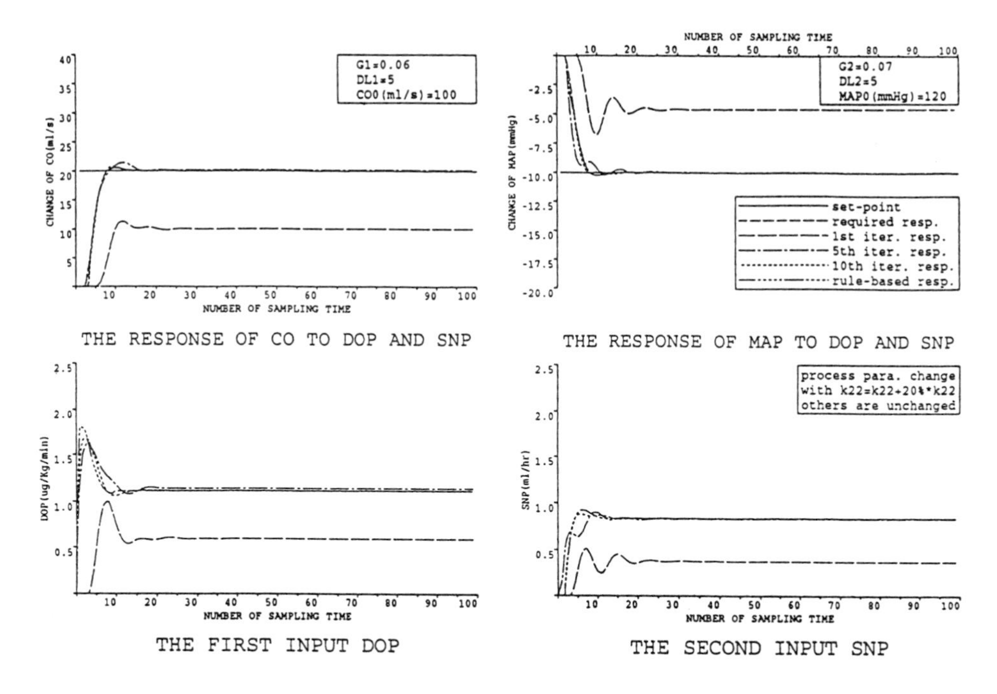


**Figure 5.7** Responses of the learning and fuzzy control: process gain $K_{22}$ changed

in Figure 5.9. Note that the three curves are comparable because they are, by definition, either absolute (for *ITAE*) or relative (for *ISE*) difference with respect to the same quantity, desired *ITAE* or *ISE*. As expected, larger gains result in a steeper curve indicating a faster convergence speed in terms of performance measures. For instance, convergence takes place more quickly in the case of $g_1 = 0.08$, $g_2 = -0.09$ than that for $g_1 = 0.04$, $g_2 = -0.05$. However, increasing the learning gains may cause the system to oscillate. Therefore, a compromise between the speed and stability is needed.

(b) *Initial control actions*

Since, during the learning process, there is no apparent basis for setting the values of initial control $u_{dop}^0(i)$ and $u_{snp}^0(i)$, $i = 1, 2, ..., L$, they were simply chosen to be zero in all the simulations undertaken. Obviously, reasonable initial control values may bring the learning process to convergency more rapidly. First, a simple rule-base derived from common control knowledge as adopted in Chapter 2 was used to produce two sets of initial control $u_{dop}^0(i)$ and $u_{snp}^0(i)$. Afterwards, the same learning procedures as used before were applied. By choosing two different scaling factors in producing the initial control, two sets of results were obtained. The results have indicated that a faster

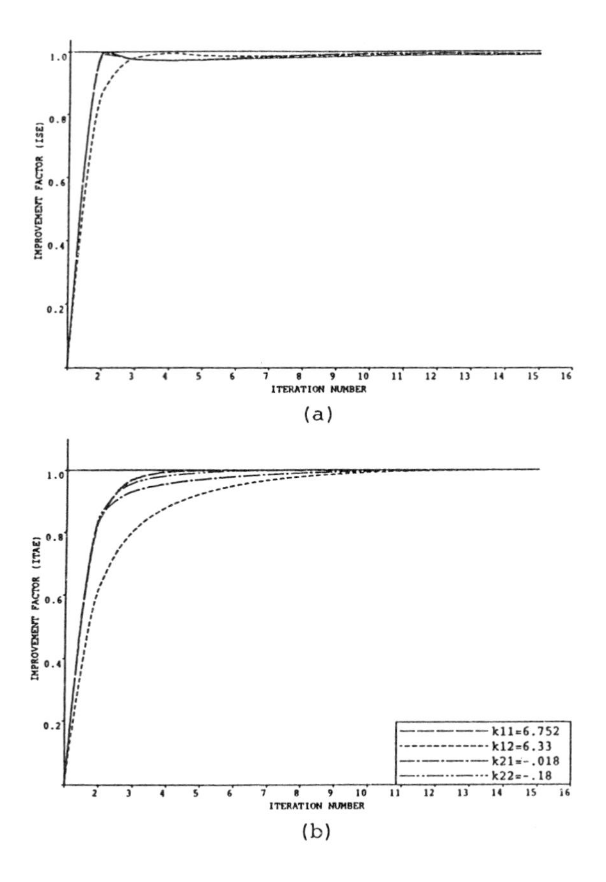


**Figure 5.8** Improvement factor *IP* for *MAP* during learning: process gains changed

convergency speed could be achieved if initial control values were appropriately chosen. This can be seen clearly by inspecting *AD* or *RD* curves as shown in Figure 5.10. Here only a detailed version of *ITAE* is presented. *AD* measures of the non-zero initial control approach zero much more quickly than those of zero initial control, suggesting that fewer iterations are needed in the former case.

(c) *Different update laws*

The above simulations involve only the error-correction update law in which the control update in each loop is based only on the learning error produced by its own loop. By setting $g_1 = 0.06$, $g_2 = 0.07$ and $Cg_1 = Cg_2 = 0.8$ in equations (5.35) and (5.37), the simulation results using the change-in-error and the cross-error correction algorithms have shown that the output responses

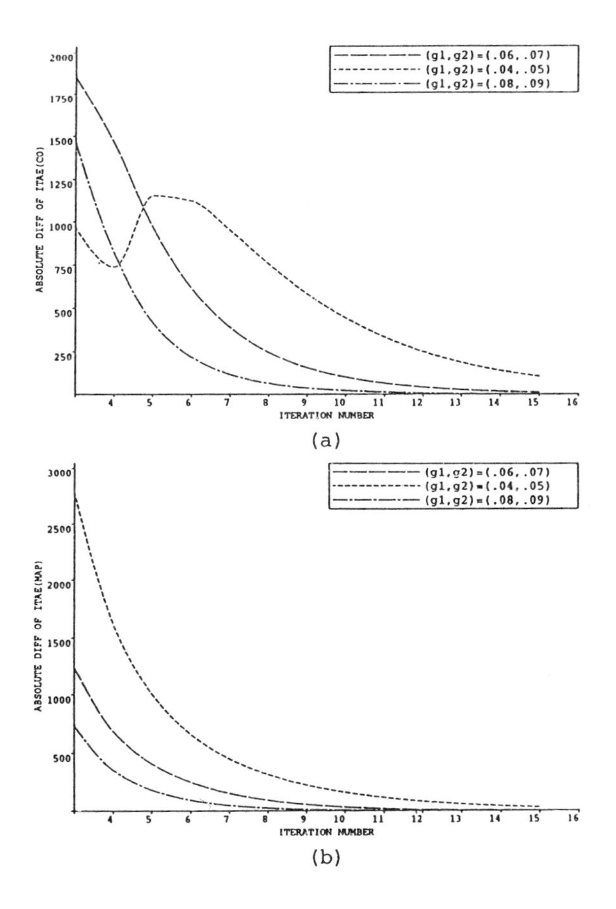


**Figure 5.9** detailed version of *AD* with different learning gains

in both cases approached the desired one with increasing iteration number.

We note that it is not easy to directly compare the learning speed within three schemes in a meaningful way because of the difficulty arising from determining the learning gains to be used for comparisons. It makes sense only if the comparisons are made under the condition that optimal gains for each algorithm are employed. Here rather than seeking the optimal gains, we simply make a comparison in the sense of contributed correcting proportion of one loop's error to the other part (change-in-error of own loop or the other loop's error). *AD* performance measures are shown in Figure 5.11, where the three curves correspond to Figure 5.5 (the proportions of 1 to 0), to the case of proportions of 0.8 to 0.2 in the change-in-error algorithm, and to that of proportions of 0.8 to 0.2 in the error-cross correction algorithm. We see that the

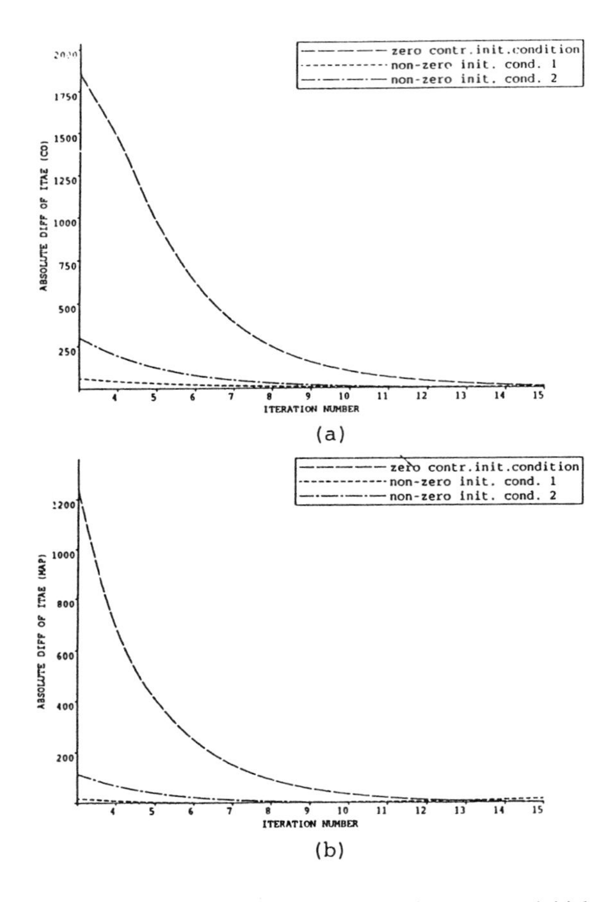


**Figure 5.10** Detailed performance measures *AD* with non-zero initial control

convergence speeds of both the cross-error and the change-in-error correction algorithms are slower than that of the error correction algorithm. It should emphasized that it is not clear whether this statement can be applied generally. However, it seems that a faster speed is achieved in the cross-error algorithm compared to the change-in-error correction algorithm. The learning ability of both algorithms with different learning gains $Cg_1$ and $Cg_2$ has also been investigated. In all the cases, convergence is assured and the convergence speed is affected by the values of the learning gains.

3. *Learning robustness*
   The term *learning robustness* is used here to indicate how robust the learning algorithm is with respect to variations in the learning gains and estimated time delays h and to the noise-contaminated measured process output.

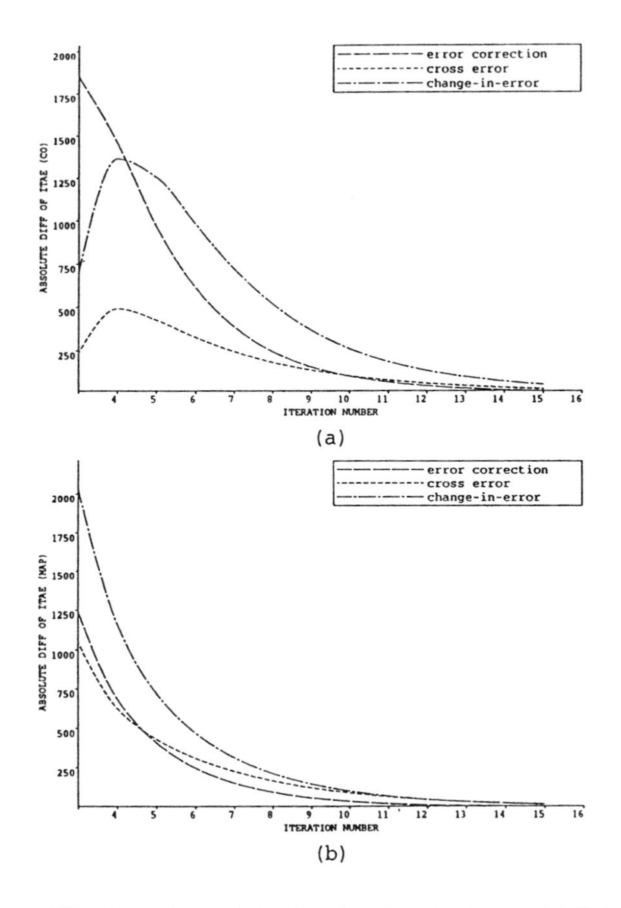


**Figure 5.11** Comparisons of the three learning algorithms: Detailed *AD*

The first group of simulations have, in fact, demonstrated that the learning algorithms are very robust with respect to the learning gains in the sense of convergence although the learning speed is affected by the values of the gains.

The estimated time delay $h$ in the learning update laws is the other important factor besides the learning gains in determining the learning performances. Recall that it has been chosen as $h = 5$ for the two loops throughout previous simulations. Denoting $h_1$ and $h_2$ as the estimated time delays for loop 1 and loop 2 respectively, the following three sets of $h$ were chosen to perform the simulations with the other conditions being exactly the same as in the case of Figure 5.5.

$$h_1 = 4 \qquad h_2 = 4$$
$$h_1 = 5 \qquad h_2 = 3$$
$$h_1 = 7 \qquad h_2 = 5$$

It has been found that relatively good performances can be achieved in the first two cases, but the convergence performance in the transient stage is relatively poor in the last case as shown in Figure 5.12. Nevertheless, the responses are near the desired one and more importantly the steady-state response is tightly convergent to the desired one. Therefore, we can conclude that the value of $h$ has a greater effect on the transient period than on the steady-state period, and the selection of $h$ is not too crucial due to the relatively robust property of the learning process.

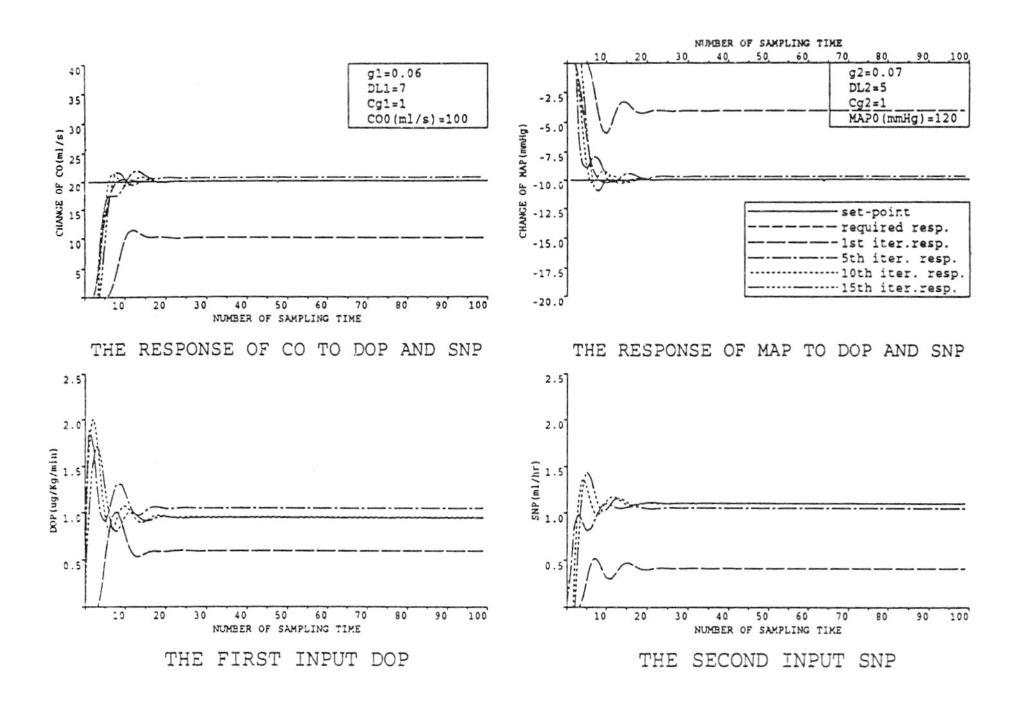


**Figure 5.12** Responses during learning with $h$ changed: $h_1 = 7$; $h_2 = 5$

Finally, we examine the effects caused by noise-contaminated data. Intuitively, the learning performance will be degraded to some extent if the measured outputs, hence the learning errors, are heavily contaminated by noise. A uniformly distributed noise signal with zero mean value and amplitude of 5% about the set-point was added to the outputs $y_{co}^k$ and $y_{map}^k$. Figure 5.13 shows the simulation result. It can be seen that convergence to the desired response within a small deviation was achieved despite the presence of the noisy learning signals.

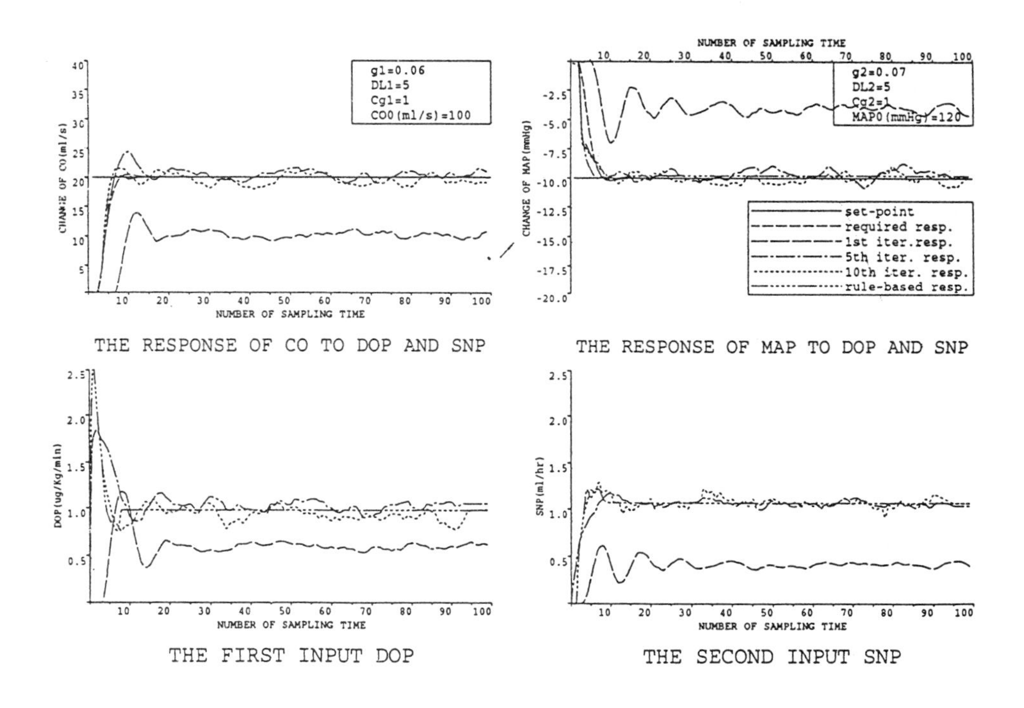


**Figure 5.13** Responses of learning and fuzzy control with noisy learning errors

Note that we deliberately let the outputs of the fuzzy controller be noise-free in order to display the ability of the proposed method to build the rule-base in a noisy environment. Although the learned control signals fluctuated in the steady-state, the fuzzy control signal remained constant. This is due to the conflict resolution procedure involved during the rule formation stage which assures that not too many rules are extracted. In fact, 18 and 16 rules were obtained for two loops in this case, which are only slightly more than those in the noise-free case.

## 5.5 Summary

Continuing from Chapter 4, this chapter presents a methodology for constructing the rule-base from learned data. The approach, based on a simplified fuzzy control model, is systematic and simple but efficient. The proposed system has been applied to the problem of multivariable control of blood pressure which is characterized by strong interactions and pure time delays in controls. It is shown that the effects of loop interaction are removed automatically by the learning scheme so that a decoupled control structure can be built. Moreover, the problem of pure time delay is tackled well due to the iterative property leading to the fact that neither state nor output prediction is needed.

A variety of simulation studies have revealed that the blood pressure system can be controlled satisfactorily and that the proposed method is feasible. Some important conclusions drawn from the simulation studies are presented below:

1. The desired transient requirement for each loop can be specified separately.
2. The system possesses a high adaptability with respect to various operating conditions such as variations of the desired response and the process parameters.
3. The reproducibility of the rule-based fuzzy control can almost always be guaranteed.
4. The learning gains and the initial control are the main factors in determining the learning speed.
5. Although all the three learning update laws can perform well, it seems that the error correction method is the best choice due to its simplicity.
6. The learning system is relatively robust with respect to variations in the estimated time delay and noisy measurement data.

CHAPTER 6

# Neural network-based approximate reasoning: principles and implementation

*Based on the principle of functional equivalence, it is demonstrated that knowledge representation and approximate reasoning can be carried out by a Back-propagation Neural Network (BNN) with the aid of fuzzy set theory.*

## 6.1 Introduction

Approximate reasoning can be regarded as a process by which a set of imprecise conclusions is deduced from a collection of imprecise premisses (Zadeh 1973). This statement, in fact, indicates two important features possessed by the method. First, the information involved in the system is linguistic, and thus, usually imprecise. Second, an approximate conclusion can be reached even though the incoming data do not match any rule's premise exactly.

Although it is well known that the method using linguistic terms to represent human knowledge possesses a high ability in terms of information fusion or abstraction, a real implementation of a specific approximate reasoning system has to rely on a nonfuzzy computer which is naturally a numerical information processing system. This is particularly true in the case of fuzzy control. This fact, qualitative description versus quantitative processing, provides a first and intuitive motivation to consider the possibility of implementing the AR system by neural networks in one form or another. However, the motivation comes mainly from the fact that the goals of AR and mapping neural networks are very similar, that is, to perform some kind of approximation or interpolation, although the representation of the information they are dealing with is different. In particular, we claim that neural networks inherently possess some fuzziness, the source of approximation in AR, which is displayed in the network via net-computing in the form of generalization. This view is of particular importance in net-based fuzzy control as will be discussed in Section 6.6.

Instead of seeking a structure mapping from a fuzzy reasoning system to a neural network, this chapter aims to find a functional mapping from the fuzzy logic-based algorithm to the network-based approach. By viewing the given rule-base as defining a global linguistic association constrained by fuzzy sets, approximate reasoning is implemented here by a Back-propagation Neural Network (BNN) with the aid of fuzzy set theory. By paying particular attention to the capability of generalization of the BNN, the underlying principles have been examined in detail using two examples: a small demonstration at the linguistic level, and a more realistic problem of multivariable fuzzy control of blood pressure. The simulation results not only indicate the feasibility of the BNN-based approach, but also reveal some deeper similarities which exist in the two methods, which may have some important implications for future studies into fuzzy control. In addition, this work may be considered as another application example of the BNN in the case of continuous outputs and on a relatively larger scale (in the second example the BNN has 26 inputs and 13 outputs, with a total of 2013 weights and thresholds). Furthermore, the work may provide evidence to support the argument that net-computing is, in fact, knowledge representation as claimed by Pao and Sobajic (1991).

In the next section, the problem we intend to deal with is formulated. Section 6.3 presents the methods for implementing reasoning functions by means of multilayered feedforward networks. Taking a small rule-base as an example, the approximation ability of the network at the linguistic level is investigated in Section 6.4. Following this, the problem of multivariable control of blood pressure is considered in Section 6.5, whereby the reasoning mechanism is performed by a network.

## 6.2 Formulation of the problem

Suppose that, for a specific problem, we have obtained a set of linguistic rules expressed in the form of IF *situation* THEN *action* statement from domain experts. Furthermore, these rules are supposed to be consistent and independent of each other, with the latter meaning that no causal relationships exist among rules. Then the problem we are interested in is how to deduce a reasonable conclusion when a specific situation, which may not be included in the IF part of any rule in the rule-base, is presented.

More specifically, assume that the system has $n$ inputs and $m$ outputs denoted by $X_1, X_2, \cdots, X_n$ and $Y_1, Y_2, \cdots, Y_m$. Furthermore, it is assumed that $L$ IF-THEN rules, and n inputs in the IF part and $m$ outputs in the THEN part are connected by linguistic connectives ALSO and AND respectively. Then the problem may be described as follows:

- Given rules: $RULE^1$ ALSO $RULE^2$ ALSO $\cdots$ ALSO $RULE^L$
  where $RULE^j$ has the form:
  IF $X_1$ is $A_1^j$ AND $X_2$ is $A_2^j$ AND $\cdots$ AND $X_n$ is $A_n^j$

THEN $Y_1$ is $B_1^j$ AND $Y_2$ is $B_2^j$ AND $\cdots$ AND $Y_m$ is $B_m^j$, $j = 1, 2, \cdots, L$.

- Given input data: $X_1$ is $C_1$ AND $X_2$ is $C_2$ AND $\cdots$ AND $X_n$ is $C_n$
- To find output data: $Y_1$ is $D_1$ AND $Y_2$ is $D_2$ AND $\cdots$ AND $Y_m$ is $D_m$ where $X_i$ and $Y_k$ are linguistic variables whose values are taken from the universes of discourse $U_i$, $V_k$ with $U_i = (u_{i1}, u_{i2}, \cdots, u_{is_i})$ and $V_k = (v_{k1}, v_{k2}, \cdots, v_{kr_k})$. $A_i^j$, $C_i$, $B_k^j$, $D_k$ are fuzzy subsets which are defined on the corresponding universes, and represent some fuzzy concepts such as *big, medium* and *small* etc. More precisely, fuzzy subsets $A_i^j$, $C_i$, $B_k^j$, $D_k$ are characterized by the corresponding membership functions $A_i^j(u_i) : U_i \rightarrow [0,1]$ , $C_i(u_i)$: $U_i \rightarrow [0,1]$ , $B_k^j(v_k)$: $V_k \rightarrow [0,1]$, and $D_k(v_k) : V_k \rightarrow [0,1]$.

To summarize, the problems of concern are how to represent knowledge described by rules numerically and how to infer an approximate action in response to a novel situation. We have tackled this problem extensively in Chapter 2 under the framework of possibility theory, thereby providing a solution using logic algorithms. In what follows, we present an alternative treatment to this problem using neural networks.

## 6.3 Solution using neural networks

### 6.3.1 BNN network

Numerous neural networks have been suggested and investigated with various topologic structure, functionality, and training algorithms (Hecht-Nielsen 1990). We are particularly interested in those which can be useful for our purpose, i.e. as mentioned previously, those which can perform functional approximations. Although the Back-propagation Neural Network (BNN), the Basis Function Network (BFN), and Probability Function Network (PFN) are good candidates, the well-known BNN structure has been chosen for the current study.

The BNN (Rumelhart *et al.* 1986) is a multi-layered network consisting of fully interconnected layers comprising many simple and identical processing units (nodes). An architecture of the BNN with one input layer, one hidden layer and one output layer is shown in Figure 6.1. Let $N_I$ inputs and $N_O$ outputs of the network be denoted as $u = [u_1, u_2, \cdots, u_{N_I}] \in R^{N_I}$ and $v = [v_1, v_2, \cdots, v_{N_O}] \in R^{N_O}$. The $i$th component $v_i^a$ of output $v^a$, when a specific input $u^a$ is presented, can be calculated by

$$v_i^a = f_o\left[\sum_{j=1}^{N_H} w_{ij}^o \cdot f_h(net_j) + \theta_i^o\right] \tag{6.1}$$

where $i = 1, 2, \cdots, N_O$,

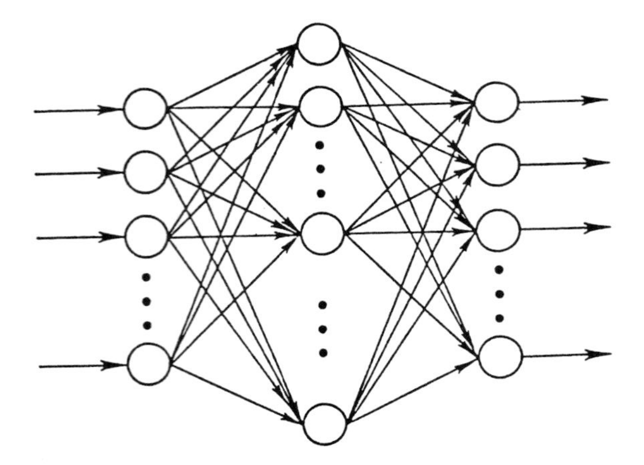

**Figure 6.1** An architecture of BNN

$$net_j = \sum_{k=1}^{N_I} w_{jk}^h \cdot u_k^a + \theta_j^h \quad (6.2)$$

and $N_H$ is the number of units in hidden layer. $w_{ij}^o$ is the connection weight connecting the $j$th unit in the hidden layer to the $i$th unit in the output layer and $\theta_i^o$ is the threshold of the $i$th unit in the output layer. $w_{jk}^h$ is the connection weight connecting the $k$th unit in the input layer to the $j$th unit in the hidden layer and $\theta_j^h$ is the threshold of the $j$th unit in the hidden layer. $f_h()$ and $f_o()$ are called activation functions which are assumed to be continuous, bounded and non-constant. Two frequently used functions are given by

$$f(net) = \frac{1}{1 + \exp(-net)} \quad (6.3)$$

and

$$f(net) = \frac{1 - \exp(-net)}{1 + \exp(-net)} \quad (6.4)$$

It should be noted that the active functions $f_o()$ at the output layer can be linear functions when continuous outputs are required.

Training in the BNN network involves adjusting the weights of the connections between the processing units under a supervised scheme. A specific training pattern $(u^p, v^p)$ is supplied to the input and the output of the network. The actual output $u^a$ corresponding to the present weights is calculated by equation (6.1) and is compared with the desired output $v^p$. Thus, the errors at the output layer are obtained and back-propagated to the individual connections in the network. The weights are updated, either after one pattern is presented or after all of patterns are entered, in the direction of steepest gradient descent in error space. The process is repeated until the performance of the network is satisfied. To be more specific, the change of the weight follows a rule called the generalized delta rule which is given by

$$\Delta w_{ji} = \eta \cdot \delta_j \cdot O_i \tag{6.5}$$

where $O_i$ denotes the input from the $i$th unit in input layer (hidden layer) to the $j$th unit in the hidden layer (output layer) and

$$\delta_j = f'(net_j) \cdot (v_j^p - v_j^a) \tag{6.6}$$

for the units in the output layer, and

$$\delta_j = f'(net_j) \cdot \sum_{k=1}^{N_O} \delta_k \cdot w_{kj} \tag{6.7}$$

for the units in the hidden layer, $f'$ is the derivative of the activation function with respect to the $net_j$, and $\eta$ is the learning rate.

### 6.3.2 Isomorphic mapping of functionality

In order to obtain a formal equivalence or isomorphic mapping between the BNN and the AR from the functional viewpoint, we formulate both of them in a unified way as described below.

From the mapping perspective, the goal of the BNN is to perform an approximate implementation of a unknown mapping $\phi$, from a compact set $I^{N_I} \subset R^{N_I}$, an $N_I$-dimensional Euclidean space, to an $N_O$-dimensional Euclidean space, $\phi$: $I^{N_I} \rightarrow R^{N_O}$, by an approximator $\phi^*$ consisting of layered and massively connected processing units. To construct $\phi^*$, the topology of the network must be determined first, then parameters concerning the specified structure are selected, and finally the performance of the resultant approximator is tested. The first step is usually taken in an *ad hoc* manner, but the method for selecting the parameters is well developed theoretically. The basic idea is to find a set of parameters for which the constructed approximator can perform best in some sense over a set of selected examplars taking from $I^{N_I}$ and $R^{N_O}$ respectively. In particular, assume that we are given a set of examplars $(u^1, v^1), \cdots, (u^P, v^P)$ with $u^p$ being drawn from $I^{N_I}$, and $v^p$ being supposed to be satisfied with the unknown function $\phi$, i.e. $v^p = \phi(u^p)$. Denote all the weights and thresholds in the structure-specified BNN as a parameter vector $w$ which is subject to being determined via training. Furthermore, the $P$ desired and $P$ actual outputs calculated from the BNN at $P$ discrete sample points are denoted as $\phi(u) \equiv [\phi(u^1), \phi(u^2), \cdots, \phi(u^P)]$ and $\phi^*(u, w) \equiv [\phi^*(u^1, w), \phi^*(u^2, w), \cdots, \phi^*(u^P, w)]$ respectively. Then the problem can be simply formulated to select $w$ in such a way that a specific error function, say a quadratic one, as denoted by

$$E(w) = \frac{1}{2} \| \phi(u) - \phi^*(u, w) \|^2 \tag{6.8}$$

is minimized, provided that the structure of the BNN is specified *a priori.*

Assuming that the unknown function is well characterized by the selected examplars, we can hope with some confidence that the approximator constructed from only a finite set of points in $I^{N_I}$ will work satisfactorily over the whole space of $I^{N_I}$. In other words, the approximator should have the ability of generalization, approximation or interpolation in response to the unseen examplars.

Based on the idea discussed above, the problem of the approximate reasoning presented in Section 6.2 may be solved if it is formulated in the same way as in the BNN, that is, to construct an inference engine $\Psi^*$ based on the given $L$ rules and to generalize to unseen situations in the domain of interest by directly manipulating the engine. Clearly, a given rule is analogous to an examplar in the BNN. $L$ IF-THEN statements may designate an implicit and global relationship between the situation set and action set. To be more specific, denote the situation variable $X = [X_1, X_2, \ldots, X_n]$ and the action variable $Y = [Y_1, Y_2, \ldots, Y_m]$. $X$ and $Y$ take linguistic labels, represented by fuzzy sets $A_i^j$ and $B_k^j$ defined on the corresponding universes $U_i$ and $V_k$, as their values, for example, $X = X^p = [A_1^p, A_2^p, \ldots, A_n^p]$. Further, let $\Psi(X) = [Y^1, Y^2, \ldots, Y^P]$ and $\Psi^*(X, W) = [\Psi^*(X^1,W), \Psi^*(X^2,W), \ldots, \Psi^*(X^P,W)]$ be the desired action and the actual action calculated from the inference engine $\Psi^*$ with respect to $P$ situations $X^p$, $p = 1, P$, where $W$ denotes a set of parameters determining the $\Psi^*$. Then, by analogy with equation (6.8), the problem may be reformulated to select $W$ in such a way that

$$\tilde{E}(W) = \frac{1}{2}\|\Psi(X) - \Psi^*(X,W)\|^2 \tag{6.9}$$

is minimized such that an action $Y = [D_1, D_2, \ldots, D_m]$ will be approximately deduced from the constructed inference engine $\Psi^*$ when a new situation $X = [C_1, C_2, \ldots, C_n]$ is encountered. Thus, the $\Psi^*$ accomplishes an approximate implimentation of linguistic mapping from one linguistic set of the situation domain to another liguistic set of the action domain. However, it should be noted that, in contrast to the point-to-point mapping in the case of $\phi^*$, the $\Psi^*$ carries out a set-to-set mapping with the constraint imposed by the corresponding membership function.

It is obvious, from the above discussions, that the AR problem can be solved by the network method if the linguistic values can be translated into numerical forms suitable for the use of the BNN and the numerical outputs from the BNN can be converted as linguistic labels. The graded membership function is the first choice for the former transformation and some methods for linguistic approximation (Eshragh and Mamdani 1981) may be used for the latter conversion. This is schematically illustrated in Figure 6.2, where the inputs and outputs of the BNN are membership grades and are bounded in the interval of [0,1]. It is noted that although the reasoning system itself has $n$ inputs and $m$ outputs, the BNN using the fuzzy set representation will have $s_1+s_2+\cdots+s_n$ inputs and $r_1+r_2+\cdots+r_m$ outputs with $s_i$ and $r_k$ being the cardinality of $U_i$ and $V_k$ on which fuzzy sets are defined, that is, $u \in [0,1]^{s_1+s_2+\cdots+s_n}$ and $v \in [0,1]^{r_1+r_2+\cdots+r_m}$. There exist other possibilities for the label/numerical value transformation, one of which will be discussed in Section 6.5.

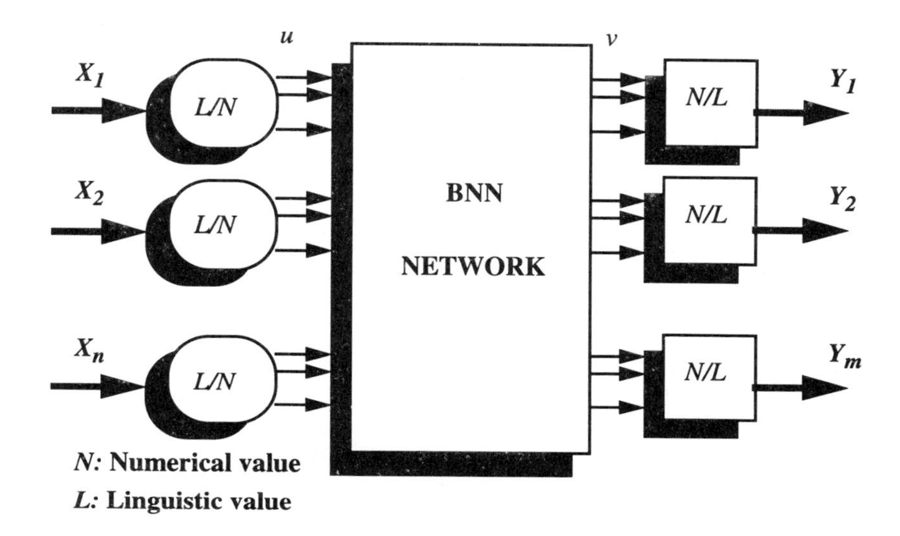


**Figure 6.2** BNN-based reasoning system

It should be pointed out that the BNN methods can also be used to handle the case where the fuzzy rule and input data may contain a probability uncertainty, an example being "IF $\{X$ is $A\}$ THEN $\{Y$ is $B$ with certainty 0.9$\}$" and "$X$ is $C$ with certainty 0.95". By changing them into "IF $X$ is $A$ THEN $Y$ is $\bar{B}$" and "$X$ is $\bar{C}$" using some modifying algorithms presented in Chapter 2, it is clear that this kind of knowledge can be coped with by the method presented previously without involving a change of net structure. The other point we would like to emphasize is that the above discussion on the equivalence between the BNN and the fuzzy system focuses only on the functional mapping aspect. The relationship between these two systems in the other aspects will be addressed in the following sections.

## 6.4 Reasoning capability: a linguistic study

Although the theoretical investigation (Hornic 1991; Hornic *et al.* 1989) into the approximation capability of the BNN has shown that the standard BNN even with one hidden layer is a universal approximator so as to guarantee the existence of the required network with respect to the approximated function, no effective method is available to guide the selection of structure parameters, for example the number of hidden units, for such a network. This is because the proof procedures are usually not constructive. In our case, we are particularly concerned with the generalizing or reasoning capacity of the BNN depending not only on the net size, but also on the attribution of the training examplars (rules in this case). We notice that Baum and Haussler (1989) have addressed this kind of question and established some relationships between the net size and the valid generalization under the case of binary outputs. However, the results do not apply to our case where continuous outputs are required.

The main objective of the study in this section is two-fold: (a) to investigate the capability of generalization of the BNN-based reasoning system at the linguistic level and (b) to examine the effect of the number of hidden units upon this capability, by means of simulation on a small but typical problem.

Suppose we are given five rules, each of which has the form of "IF $X$ is $A$ THEN $Y$ is $B$" with the pair $(A, B)$ being specified as $(PB, NB)$, $(PM, NM)$, $(ZR, ZR)$, $(NM, PM)$ and $(NB, PB)$, where $PB$, $PM$, $ZR$, $NM$, and $NB$ stand for Positive-Big, Positive-Medium, Zero, Negative-Medium and Negative-Big respectively. For simplicity, we assume that all the fuzzy sets representing the above linguistic labels are defined on the same universes of discourse, i.e. $U = V = \{-4, -3, -2, -1, 0, 1, 2, 3, 4\}$ and all the membership functions are of triangular form centrally located at –4, –2, 0, 2, 4 with an identical width of 3. The BNN consists of one input and one output layer with nine input and nine output units, corresponding to the cardinalities of $U$ and $V$, and one hidden layer with a variable number of units. By setting the initial weights to be uniformly distributed on [–0.5, 0.5], the net was trained by the presentation of five rules with the learning rate being 0.8 and the number of hidden units being 5, 15, 30 and 50 respectively. The training process was stopped when the sum of squared error within one cycle was less than 0.0005. Then the BNN was tested using the following two sets of linguistic labels.

The first set of linguistic inputs used so-called linguistic hedges to modify the basic labels in the rules. Here two hedges, *very* and *more or less* were used and defined as *very* $A \equiv A^2$ and *more or less* $A \equiv A^{0.5}$. Thus, ten different inputs could be used to test the performance. For example, if very-positive-medium is presented to the BNN, the expected outputs should be *very-negative-medium*. Figure 6.3 shows four such cases, each consisting of five curves: input, expected output, and three different actual outputs corresponding to 5, 15 and 30 hidden units. It can be seen that the outputs of the BNN compare well with the expected ones, indicating a valid generalization. By comparing the results obtained with different hidden units, it may be concluded that the generalization performance of the BNN is improved with an increase in the number of the hidden units. However, this is not always true, as will be discussed below.

Instead of merely altering the shapes of the basic labels as above, the second set of linguistic inputs was concerned with changing the central values with respect to the basic labels, representing a much harder situation than the previous one since the inputs are substantially different from that in the given rules. It is convenient to interpret the linguistic labels as fuzzy numbers meaning linguistically that "$X$ is about $u$". With the same trained BNN, Figure 6.4 gives the results corresponding to

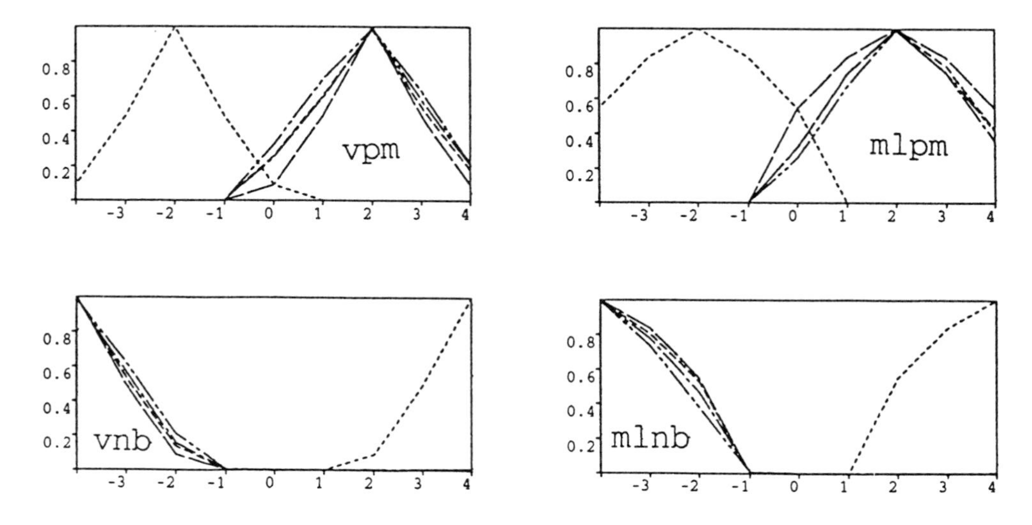


**Figure 6.3** The linguistic inputs "very" and "more or less"

the expected outputs of "about" +3, +1, −1, −3. At first sight, one may think that the results are not as good as would be expected. However, if we calculate the area under the different curves and accept this quantity as a global measure of the performance, the results are remarkably good. For example, when the expected output is "about −1", the absolute differences between the expected and actual areas (approximately calculated using a weighted sum $\sum_i u_i A(u_i)$) are 0.1552, 0.1141, 0.1142, and 0.0534 corresponding to 5, 15, 30, and 50 hidden units respectively.

It should be noted that the generalization performance of the BNN does not monotonically increase with an increase in the number of the hidden units, a problem known as overfitting. To clarify this point, we adopt another global measure, namely the centre of gravity (COG) which is an important quantity in fuzzy control, where it is used to produce a nonfuzzy output. The COG is defined by

$$COG = \frac{\sum_i u_i \cdot A(u_i)}{\sum_i A(u_i)} \tag{6.10}$$

For the present problem, the averaged differences between the expected and actual COG were 0.1239, 0.0912, 0.1109, and 0.0965 corresponding to 5, 15, 30, 50 hidden units respectively, indicating that, at least in the current context, the reasoning performance could be degraded with too few or too many hidden units.

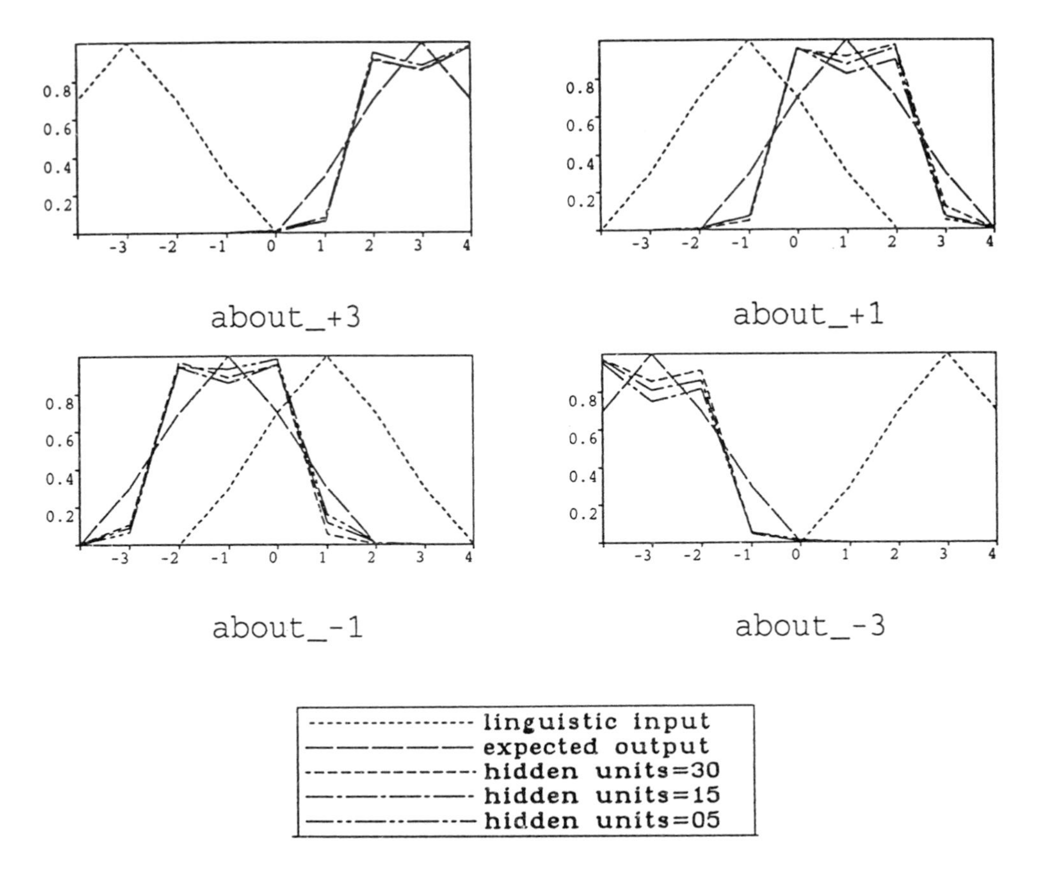


**Figure 6.4** The linguistic inputs of "about"

## 6.5 Reasoning capability: a fuzzy control example

This section presents a more complicated and realistic problem of multivariable blood pressure control solved by the BNN-based reasoning method. The control system having the same structure as the one used in Chapter 3 consists of two separated control loops, *CO*/DOP and *MAP*/SNP, with the aid of a simple compensator to reduce the interactive effects. For simplicity, the two controllers, each comprising two inputs, error $e$ and change-in-error $ce$, and one output $u$, are assumed to be identical and therefore, only one of them is described.

Suppose that all the universes of discourse have the same form consisting of 13 integers ranged from –6 to 0 to 6, and that seven linguistic labels (negative-big (medium, small), zero, and positive_big (medium, small)) are used. They are defined by the same triangular form with the central values located at −6, −4, −2, 0, 2, 4, 6 respectively and with an identical width of 3. Thus, the resultant BNN has 26 inputs and 13 outputs. If one hidden layer with 50 units is used, the BNN will have a total of 2013 adjustable parameters.

Two phases, off-line training and on-line application, are needed. Using the same rule-base as used in Chapter 3 consisting of 33 rules as shown in Table 6.1, the BNN was trained with 50 hidden units and a learning rate of 0.02. Although the actual outputs of the BNN are in the interval [0,1] in accordance with the range of a sigmoid nonlinear activation function, it has been found that a faster convergency could be achieved if linear activation functions in the output layer are used.

**Table 6.1** Control rules

| *U* | | **Change in error (*C*)** | | | | | | |
|---|---|---|---|---|---|---|---|---|
| | | **NB** | **NM** | **NS** | **ZR** | **PS** | **PM** | **PB** |
| **Error (*E*)** | **NB** | | NB | | NM | NM | NS | |
| | **NM** | NB | | NM | | NS | | PS |
| | **NS** | | NM * | NS | NS * | ZR * | PS * | PM |
| | **ZR** | NB | | NS * | ZR * | PS | | PB |
| | **PS** | NM | NS * | ZR * | PS * | PS * | PM * | |
| | **PM** | NS * | | PS * | | PM * | | PB |
| | **PB** | | PS * | PM | PM * | | PB | |

*: 16 rules

This is because the change of the weight is proportional to the derivative of the activation function, which is non-constant for a nonlinear function and is particularly in favour of only the output values around 0.5 (in the case of a sigmoid function). Obviously, this is inadequate for the case of continuous outputs where all output values are equally important. This point is clearly illustrated in Figure 6.5 by comparing the corresponding results. To investigate the effects of the number of rules upon the reasoning performance, a set of 16 rules selected from 33 rules as marked with "*" in Table 6.1 was also trained. As expected, the sum of the squared error decreases more quickly as shown in Figure 6.5, resulting from the fact that the rule-base with fewer rules would specify a less restricted global association and would be easier to learn.

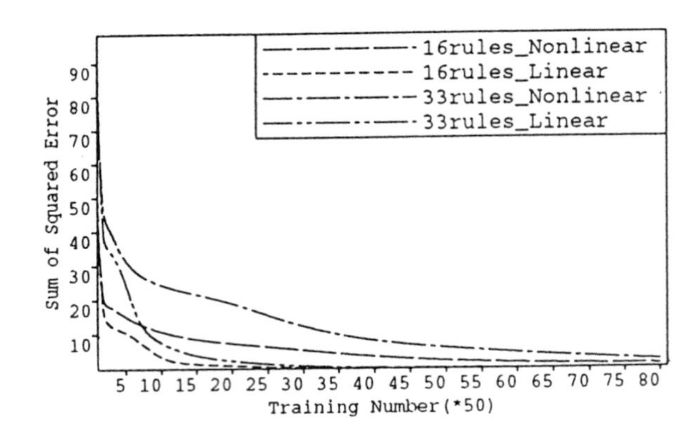


**Figure 6.5** SSE versus training cycles

The measured error $e$, change-in-error $ce$ and the required control $u$ are numerical, the $e$ and $ce$ have to be fuzzified into fuzzy sets and the fuzzy sets output of the BNN must be defuzzified into a real number during the application stage. Here singleton fuzzification and COG defuzzification methods were employed. However, if we think of $e$ and $ce$ and $u$ as the inputs and output of the controller, nothing is "fuzzy" at all! But why is it fuzzy? This is simply because the linguistic labels in the rules are interpreted as fuzzy sets characterized by the membership function and the BNN is trained accordingly. In other words, the BNN trained in a linguistic environment must be used in a numerical environment. However, if the linguistic labels are viewed as fuzzy numbers and the BNN is trained with the central values of the fuzzy numbers only, things will be totally different. The complexity of the controller is greatly reduced and the efficiency in speed and storage is dramatically increased due to the fact that the number of inputs and output is reduced to two and one and neither fuzzification nor defuzzification is needed, implying that an identical environment for training and application is established. But where is fuzzy now? We speculate that the role of fuzziness, i.e. interpolation or approximation, has been replaced in some degree by the generalization property of the BNN due to the distributiveness of the BNN. To verify this conjuncture, the BNN with 15 hidden units was trained and tested using 33 and 16 rules respectively. The results are reported next. However, we should notice that the mechanisms of generalization in these two systems are substantially different. The monotonic logistic activation functions in the BNN are responsible for generalization, enabling many hidden units to contribute to a single output. In contrast, the localized membership functions in the fuzzy system are the main sources for generalization, activating finite rules to respond to a current input. Chapters 9 and 10 will explore several neural networks whose generalizing mechanisms are similar to those of fuzzy systems.

After appropriate training, the BNN was used as the controller in the system. The capability of the generalization of the BNN was evaluated indirectly by the control performances measured by two frequently used indices: integral of square of the error (*ISE*) and integral of time and absolute error product (*ITAE*) for each output. We considered four cases: the BNN's with linear activation functions in the output layer trained by fuzzy sets with 33 and 16 rules (denoted by FS_33 and FS_16) and by fuzzy numbers with 33 and 16 rules (FN_33 and FN_16). Table 6.2 shows the simulation results when the system was subject to square-like inputs. The results obtained using the logic-based method in Chapter 3 are also included in Table 6.2 for the purpose of comparison.

**Table 6.2** Control performances

| | CISE * | CITAE | PISE * | PITAE |
|---|---|---|---|---|
| FS_33 | 16.80 | 843.47 | 3.55 | 586.24 |
| FS_16 | 16.83 | 902.16 | 3.54 | 609.35 |
| FN_33 | 16.79 | 810.29 | 3.49 | 560.72 |
| FN_16 | 16.78 | 810.27 | 3.50 | 558.87 |
| MMC | 16.77 | 808.43 | 3.48 | 557.06 |

* C for *CO* and P for *MAP*

The following conclusions can be drawn from the results: (a) the BNN possesses a high ability to generalize, a very useful property for fuzzy reasoning. This is particularly demonstrated by the case of FS_33 and FS_16 where the measured singleton inputs were very different from the training inputs; (b) it is possible to use fewer rules to obtain a valid generalization as illustrated by FS_16 and FN_16; (c) perhaps the most important conclusion drawn from the results is that the BNN trained by the fuzzy numbers can produce performances as good as, or even better than, the one trained by the fuzzy sets as indicated by the case of FS_33 wrt. FN_33 or FS_16 wrt. FN_16; (d) the similar performances obtained by the logic-based and the BNN-based reasoning approaches suggest the feasibility of the BNN for the application of fuzzy control.

## 6.6 Summary

The fact that a fuzzy logical problem can be solved well by means of neural-net-computing may imply some interesting and useful points. Here we will briefly discuss a few of them which are of particular interest.

*Fuzziness versus distributiveness.* By paying particular attention to the capability of generalization of the BNN, it has been demonstrated that a forward-chaining fuzzy reasoning system with parallel rule-bases can be implemented within the framework of neural networks. The studies into the BNN-based fuzzy controller suggest that, besides a seeming resemblance between rules and patterns in the logic-based and BNN-based approaches, there exists a deeper similarity in the information processing aspect of them, namely, fuzziness versus distributiveness. The terms fuzziness and distributiveness are used to characterize the nature of generalization to unseen situations, that is, fuzziness in the fuzzy system due to the graded membership functions and the distributiveness in the BNN due to the logistic activation functions are respective sources for generalization. In particular, it is indicated that the neural networks inherently possess some fuzziness which may come mainly from the distributiveness of the network. We believe that this suggestion has some important implications not only for the development of fuzzy control under the framework of neural networks, especially if the learning ability of the neural networks is particularly taken into account, but also for the further clear understanding of the behaviour of two different systems in a unified way. These points will be addressed further in the following two chapters where the major effort is devoted to developing BNN-based fuzzy control systems with self-learning capability.

*Knowledge representation.* Knowledge traditionally represented by the rules in rule-based expert systems is now transferred into the network via training, where the knowledge is implicitly and distributively represented by the connection weights and local processing units. Retrieval of the encoded knowledge can be carried out by directly manipulating the network. It is in this sense that we may say that knowledge can be represented by the training and computing. In other words, the training plus the computing is knowledge representation. A similar point has been emphasized very recently by Pao and Sobajic (1991).

*Knowledge processing.* Another more important point is that the net-computing is knowledge extension too! This means that the trained network is not only capable of memorizing the original knowledge, but capable of extending it to a considerable variety of new situations. It does this via processing encoded knowledge by implicitly employing meta-knowledge, i.e doing similar things when facing similar circumstances. This characteristic is of particular importance for many applications such as approximate reasoning. It is evident that the simulation results reported in this chapter provide evidence to support these points.

CHAPTER 7

# BNN network-based fuzzy controller with self-learning teacher

*A multi-stage mechanism which is able to extract teacher signals for use with a BNN-based fuzzy controller is described. Three detailed structures for the controller implementation are discussed.*

## 7.1 Introduction

By considering the previous chapter as a first step towards integrating rule-based fuzzy controllers with neural networks from a viewpoint of functional equivalence, this chapter continues the process by making a crucial assumption that no control experts are available for the particular control problem and therefore, the control rule-base must be constructed from controlled environment directly. The main interests of this chapter lie in constructing a BNN-based fuzzy controller automatically with the following requirements: (1) the controller serves as a direct feedback controller in a closed-loop; (2) the controller is able to deal with multivariable processes with possible pure time delays; (3) the whole system satisfies arbitrary but achievable performance requirements; (4) little prior knowledge about the process is assumed; and (5) the method should be as simple as possible.

By examining the existing schemes outlined in Chapter 1, it appears that none of them is adequate to meet the above requirements for various reasons. For example, the methodologies suggested by fuzzy control researchers are unsuitable because we use the net structure instead of fuzzy reasoning algorithms. The *critic* scheme may be efficient in controlling a cart-pole system where the only requirement is to prevent the pole from falling, but it is inadequate for handling tracking and regulation problems. The inverse method suffers from the limitation of open-loop applications and from the inflexibility of meeting arbitrary performance requirements. The back-propagation through identified model may be a systematic and natural strategy. However, the algorithm is relatively complicated and the controller built in this way cannot provide explicit control rules.

Taking the preceding discussions into account, we present a general mechanism capable of extracting training rules automatically from the process being controlled for use with BNN-based fuzzy controllers by incorporating a learning algorithm proposed in Chapter 3 into the control systems. With respect to the input types used by the controllers, we propose three possible structures for BNN-based fuzzy controllers.

The chapter is organized as follows. In Section 7.2 the basic principles and detailed implementation of BNN-based fuzzy controllers are presented. The overall system structure concerning the learning and rule extracting are described in Section 7.3. The simulation study using proposed methods is reported next. Some relevant points are discussed in Section 7.5.

## 7.2 BNN-based fuzzy controller

As is well known, two key issues concerning the implementation of the fuzzy controller are knowledge representation and associated inferencing schemes. Traditionally, they are carried out from a logical point of view by performing some fuzzy logic calculations, thereby facilitating a programmed implementation. In the previous chapter, we have made an attempt to implement the approximate reasoning by means of Back-propagation Neural Networks (BNN). The underlying principles lie in the fact that by viewing the given rule-base as defining a global linguistic association constrained by fuzzy sets, a functional mapping from the fuzzy logic-based algorithm to the network-based approach can be established and approximate reasoning is implemented by a BNN with the aid of the fuzzy set theory.

A block diagram of the proposed BNN-based fuzzy control system is depicted in Figure 7.1. It is a feedback control system with a similar structure to that for a traditional fuzzy control system. However, the system here works in two distinct modes: training and application. Assuming that the rule-base is given, the BNN network is trained off-line by presenting all rules sequentially to the network. Then the successfully trained network (in terms of having a sufficient accuracy) can be inserted in the control loop for on-line operation.

Depending on the methods for converting qualitative/linguistic labels into quantitative/numerical values, the structures of the resulting controllers are significantly different. There are at least two possibilities for doing this, namely, fuzzy set interpretation characterized by graded membership functions and fuzzy number translation featured typically by central values and spread widths. These are considered in the following sections.

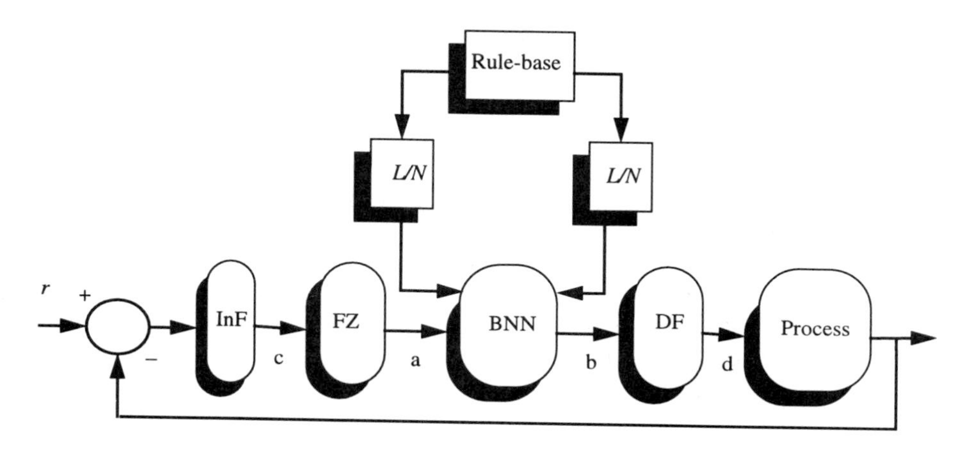


**Figure 7.1** A diagram of BNN-based fuzzy control system

### 7.2.1 Fuzzy set representation

Referring to Figure 7.1, the functions of each component in the figure under the representation of fuzzy sets will be explained briefly.

*Process.* Throughout this chapter, we will make the following assumptions about the process being controlled: (1) it is a $m$-input $m$-output process; (2) there exist significant interactions between variables and possible pure time delays in control inputs; (3) the process is not necessarily linear but the input–output characteristics are monotonic with known signs; (4) the process is open-loop stable. It is noted that the assumptions are quite mild and practical.

*Input formation.* The function of this block is to convert a $m$-dimensional control error vector $e_c$ into a $2m$- or $3m$-dimensional controller input vector. More formally, suppose that the measured control error at time $lT_s$ with $T_s$ sampling period is denoted as $e_c(lT_s)$=[$e_{c1}(lT_s)$, $e_{c2}(lT_s)$, $\cdots$, $e_{cn}(lT_s)$]. Each component of $e_c(lT_s)$ is extended into three elements: $e_c(lT_s)$, $c_c(lT_s)$, and $s_c(lT_s)$, where

$$c_{cl}(lT_s) = e_c(lT_s) - e_c((l-1)T_s) \tag{7.1}$$

is the change-in-error and

$$s_c(lT_s) = \sum_{i=1}^{l} e_c(iT_s) \tag{7.2}$$

is the sum of error. Thus, in light of different combinations, three possible controller input modes denoted by the vector $\bar{u}_0$ can be constructed, i.e

$$\bar{u}_0 = [e_{c1}, c_{c1}, e_{c2}, c_{c2}, \cdots, e_{cm}, c_{cm}] \tag{7.3}$$

$$\bar{u}_0 = [e_{c1}, s_{c1}, e_{c2}, s_{c2}, \cdots, e_{cm}, s_{cm}] \tag{7.4}$$

$$\bar{u}_0 = [e_{c1}, c_{c1}, s_{c1}, e_{c2}, c_{c2}, s_{c2}, \cdots, e_{cm}, c_{cm}, s_{cm}] \tag{7.5}$$

We call the input types determined by equations (7.3), (7.4) and (7.5) EC, ES, and ECS respectively. It is noted that these are analogues to classical PD, PI and PID controllers. In what follows, it is assumed that $\bar{u}_0$ is a $n$ dimensional vector.

*Scaling and fuzzification.* The extended vector $\bar{u}_0$ must be transferred into a new vector $u_0$ by multiplication with suitable scaling factors $g_i$, and then quantizing the scaled values to the closest elements of the corresponding universes, i.e $u_0 = [u_{01}, u_{02}, \cdots, u_{0n}]$ with $u_{0i} = [g_i \cdot \bar{u}_{0i}]$ and $u_{0i} \in U_i$. It is noted that the inputs and the outputs of the BNN shown in Figure 7.1 are assumed to be fuzzy sets. Thus, a crisp input $u_{0i}$ must be converted to a fuzzy set $\tilde{X}$. Here $u_{0i}$ is fuzzified as a fuzzy singleton $\tilde{X}_i$ with membership function given by

$$\tilde{X}_i(u_i) = \begin{cases} 1 & u_i = u_{0i} \\ 0 & u_i \neq u_{0i} \end{cases} \tag{7.6}$$

where $i = 1, 2, \cdots, n$.

*BNN network.* A BNN (Rumelhart *et al.* 1986) is typically a layered and fully connected network consisting of many identical processing units. To specify a network topology, one has to determine the number of layers, the number of units in each layer and the types of activation functions. Throughout this and the next chapters, a three-layered network, constituting an input, an output and a hidden layer, is chosen for the reason mentioned in the previous chapter. The number of the units in the input and output layers, denoted by $N_I$ and $N_O$, is determined by the total number of elements in the input and output universes $U_i$ and $V_k$ respectively, that is, $N_I = s_1 + s_2 + \ldots + s_n$ and $N_O = r_1 + r_2 + \ldots + r_m$ with $s_i = \mathrm{Card}(U_i)$ and $r_k = \mathrm{Card}(V_k)$. Although the values of inputs and outputs of the BNN are within the range [0,1], experience has shown that a network with linear activation functions at the output layer exhibits a much faster training speed than that with nonlinear activation functions. However, the activation functions at the hidden layer should be nonlinear and here sigmoid functions were chosen. So far the only undetermined parameter concerning the topology of the network is the number of hidden units. It has shown that this quantity is very important in determining both the training and generalization performances. Too many or too few hidden units will typically ledd to either a poor fitting with respect to training rules or a poor generalization with respect to new situations. While no theoretical guidance on this matter exists, it can be determined by trial and error.

As already mentioned, the BNN shown in Figure 7.1 works in two modes. During the operational mode, its inputs come directly from measured inputs with some preprocessing procedures described above, while in the training mode all the linguistic labels $ILB_i^j$ and $OLB_k^j$ in both IF and THEN parts are conceptually interpreted as fuzzy sets $A_i^j$ and $B_k^j$, which are numerically characterized by the corresponding membership functions $A_i^j(u_i)$:→[0,1] and $B_k^j(v_k)$:→[0,1]. Generally either triangular or bell-typed membership functions work well in practice.

*Defuzzification.* $N_O$ outputs from the BNN are grouped into $m$ groups, each of which is interpreted as specifying a fuzzy set defined on the corresponding universes $V_k$. However, these $r_k$ values in each group must be transferred into a determined value since the control input to the process must be crisp. Thus, a procedure of defuzzification is needed. Here, we adopt the following algorithm usually referred to as Centre of Gravity (COG) and given by

$$v_k^* = \frac{\sum_{q=1}^{r_k} \tilde{Y}_k(v_{kq}) \cdot v_{kq}}{\sum_{q=1}^{r_k} \tilde{Y}_k(v_{kq})} \qquad k = 1, 2, \cdots, m \tag{7.7}$$

where $v_k^* \in V_k$ is the defuzzified output and $\tilde{Y}_k(v_{kq})$ are the current outputs of the BNN in the $k$th group. $v_k^*$ may be multiplied by another scaling factor so as to derive an appropriate control value.

### 7.2.2 Fuzzy number representation

By carefully inspecting the method described above, we have observed the following important fact: the environments the BNN is working in for the training and operational modes are very different in terms of signal types it deals with. It is because the linguistic labels are interpreted as fuzzy sets in the training mode that the measured crisp controller inputs in operational mode must be fuzzified and the outputs of the BNN must be defuzzified so as to establish a compatible operating environment for use of both modes. An immediate effect of this is that the size of the resulting neural network will be large depending on the number of process inputs–outputs, the number of elements in the corresponding universes, and also the number of hidden units. For example, for a $2 \times 2$ process with the ECS type and with 13 elements in each of the universes, there will be $2 \times 3 \times 13 = 78$ input units and $2 \times 13 = 26$ output units in the BNN. Obviously such a realization would be extremely inefficient, in terms of training time and computer storage, and even impossible due to the BNN's limited tractability of dealing with a large scale problem. One remedy would be to adopt a decentralized strategy in which $m$ controlled variables are controlled by $m$ independent BNN-based controllers.

However, if we are able to make the environment of the training mode adapt to that of the operational mode instead of vice versa as done before, the situation would be changed dramatically. This is possible because if we think of "c" and "d" as the input and output points of the controller as indicated in Figure 7.1, then no "fuzzy" signals are involved, but rather they are completely determined. Thus, what we need to do is to enable the BNN to work at the signal level instead of the graded membership level. This can be done by interpreting the linguistic labels via fuzzy numbers as described below.

A fuzzy number is typically characterized by a central value with an interval around the centre. The central value is most representative of the fuzzy number and the width of the associated interval determines the degree of fuzziness, a wider interval indicating more fuzziness. In this perspective, the underlying realization of the BNN-based fuzzy controller can be restated as follows. By explicitly embedding the meaning of the linguistic labels, the control rule can be rewritten as

$$IF \;\; X_1 \;\; is \;\; (\hat{u}_1^j \,, \delta_1^j) \;\; AND \;\; X_2 \;\; is \;\; (\hat{u}_2^j \,, \delta_2^j) \;\; \cdots \;\; X_n \;\; is \;\; (\hat{u}_n^j \,, \delta_n^j)$$

$$THEN \;\; Y_1 \;\; is \;\; (\hat{v}_1^j \,, \gamma_1^j) \;\; AND \;\; Y_2 \;\; is \;\; (\hat{v}_2^j \,, \gamma_2^j) \;\; \cdots \;\; Y_m \;\; is \;\; (\hat{v}_m^j \,, \gamma_m^j)$$

or more compactly

$$IF \quad X \quad is \quad (\hat{u}^j \,, \delta^j) \quad THEN \quad Y \quad is \quad (\hat{v}^j \,, \gamma^j) \tag{7.8}$$

where $\hat{u}^j = [\hat{u}_1^j, \hat{u}_2^j, \cdots, \hat{u}_n^j]$, $\hat{u}_i^j \in U_i$, and $\hat{v}^j = [\hat{v}_1^j, \hat{v}_2^j, \cdots, \hat{v}_m^j]$, $\hat{v}_k^j \in V_k$ are the input and output central value vectors in the jth rule, $\delta^j = [\delta_1^j, \delta_2^j, \cdots, \delta_n^j]$ and $\gamma^j = [\gamma_1^j, \gamma_2^j, \cdots, \gamma_m^j]$ are the input and output width vectors in the $j$th rule. Now it is possible to train the BNN with only the central value vectors, i.e. $(\hat{u}^j \,, \hat{v}^j)$ pairs while leaving the width vectors untreated explicitly. One may ask the question of how the fuzzy concept is handled in such a paradigm. The answer is, as concluded in the previous chapter, that the BNN network inherently possesses some fuzziness which is exhibited in the form of interpolation over new situations. This desirable property obtained primarily from the distributiveness of the BNN is precisely the one an approximate reasoning system seeks to achieve.

Once the functional equivalence between fuzziness and distributiveness is acknowledged, the structural complexity of the BNN can be reduced greatly. This can be attributed to the fact that both fuzzification and defuzzification procedures are no longer necessary, and more importantly the number of input and output units are dramatically decreased to being equal to $n$ and $m$ respectively. Taking the same example as before, only six input units and two output units are needed in this case compared with that of 78 and 26 in the previous situation. Needless to say, the computational efficience would be greatly increased in both training and operational stages, and more importantly this interpretation provides a basis for learning and extracting rules directly from the controlled environment as described in the next section. It is worth noting that the values of the inputs and the outputs of the BNN are no longer bounded in [0,1], but are within the range of the corresponding universes. In what follows, we will only deal with this type of BNN-based fuzzy control system.

## 7.3 Learning and rule-extracting

As mentioned previously, our principal objective in this chapter is to develop a BNN-based fuzzy control system not only satisfying the desired control performances, but also requiring only little prior knowledge about the controlled process.

A difficulty arises immediately for implementation of such a controller when further assuming that the control rules or teacher signals are unavailable. Obviously, a procedure for deriving the needed rules must be determined. Here we propose a general scheme capable of achieving the above objectives and requirements. By introducing a learning mechanism, the required teacher signals are obtained first and then in turn processed to provide extraction of a set of samples for a specific controller structure which can be used to train the BNN-based controller. Thus, the whole process consists of four stages: on-line learning, off-line extracting, off-line training, and on-line operation. This cycle may be repeated if necessary. It is worth pointing out that term "learning" in this study is not exchangeable with "training" as used in the neural network literature. The details concerning the learning and extracting processes are described below.

### 7.3.1 Learning

Figure 7.2 shows a block diagram of the learning system consisting of a reference model, a learning mechanism, a controller input formation block, a short-term memory (STM), and the controlled process. By operating the learning mechanism iteratively, the desired control performances specified by the reference model subject to the command signals are gradually achieved and the desired control trajectory denoted by $v_d$ is obtained. At the same time, the controller input vectors $\bar{u}$ corresponding to a prespecified control mode EC, ES, or ECS are constructed using the procedure described in the previous subsection. The formed input vectors $\bar{u}$ together with the learned control $v_d$ are stored in the STM in ordered pairs, and are subsequently used as the basis for extracting training rules. It should be noted that there are two types of errors in Figure 7.2, namely, learning error $e_L$ and control error $e_c$. The former is defined as the difference between the output of the reference model $y_d$ and the output of the process $y_p$, whereas the latter is the discrepancy between the output of the command signal $r$ and the output of the process $y_p$.

*Performance designation.* As mentioned previously, the desired control performance is designated by the reference model specifying what the process responses should be when both of them are subject to the same command signal. It is desirable that the model should be as simple as possible so that its output response can be determined readily by adjusting a few parameters in the model. The other requirement in designing the model is that the desired performances should be achievable. For example, it is completely unrealistic to demand an instant response for a dynamic process with pure time delay subject to a step command.

Taking the requirement for simplicity into account, we propose the following reference model which is given by

$$H_r(s) = Diag\left\{H_{r1}(s),\ H_{r2}(s),\ \cdots,\ H_{rm}(s)\right\} \tag{7.9}$$

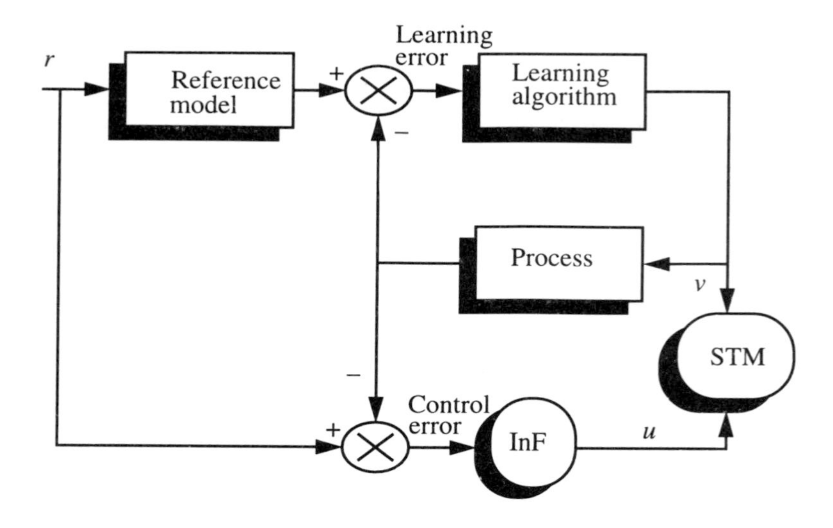


**Figure 7.2** Learning mode

where $H_r(s)$ is a diagonal transfer matrix relating the command signal $r$ to the model output $y_d$ in the $s$-domain. The model given above implies that it is linear and non-interacting between variables. It should be emphasized that the non-interacting and linear assumptions in the model do not suggest that the process itself must have these properties. Yet they do suggest that some knowledge about the process should be available. For example, we implicitly assume that the each process output is primarily affected by one of the $m$ inputs so that $m$ input and output are paired in the model.

Some advantages are obtained by using this kind of model. Being non-interacting, the control performance of each controlled variable can be specified independently without taking the other loops into consideration. This is of practical importance in some cases where, for instance, the response speed for some variables should be faster than others. The linear assumption in the model allows one to select low-order linear models which are well understood with respect to the model's response and its parameters. The following second-order model may be proposed and is given by

$$H_{ri}(s) = \frac{\omega_{nri}^2 e^{-\tau_{ri} s}}{s^2 + 2\xi_{ri}\,\omega_{nri}\, s + \omega_{nri}^2} \tag{7.10}$$

With some prior knowledge about the process and well-known linear control theory, it is not difficult to determine the parameters in the above model by specifying the desired time domain indices or by allocating the pole positions in the $s$-

plane.

*Learning algorithm.* Once the desired response $y_d$ is specified, the next step is to determine the process input $v_d$ with which the corresponding response $y_p$ of the controlled process will be close enough to $y_d$. With the assumption of little prior knowledge about the process being available, obtaining such a $v_d$ is by no means trivial. Here we adopt a simple but efficient approach usually referred to as iterative learning control as described in Chapter 3. The basic principle and a general learning algorithm are briefly outlined below.

Like any learning or adaptive system, the first concern is the learning objective which is the starting point in deriving the learning algorithm. The goal of the present learning system may be stated as follows. It is intended to force the learning error $e_L(\text{t})$ asymptotically to zero or to a predefined tolerant region $\varepsilon$ within a time interval of interest by repeatedly operating the system. More specifically, we require that $\|e_{Lk}(\text{t})\| \rightarrow 0$ or $\|e_{Lk}(\text{t})\| < \varepsilon$ uniformly in $t \in [0,T]$ as $k \rightarrow \infty$, where $k$ denotes the iteration number, $\varepsilon > 0$ is a predefined error tolerance and $\|\cdot\|$ stands for norm. Thus, we see that convergence in this case occurs along the spatial direction instead of the temporal direction as usually occurs in traditional adaptive control systems. Whenever the convergence occurred, the corresponding control action at that iteration is regarded as the learned control $v_d$. In addition to learning $v_d$, the system will also learn appropriate scaling factors for use in controller input formation.

According to the learning goal described above, the corresponding learning law is given by

$$v_{k+1}(t) = v_k(t) + P_L \cdot e_{Lk}(t+\lambda) + Q_L \cdot \dot{e}_{Lk}(t+\lambda) \tag{7.11}$$

where $v_{k+1}, v_k \in R^m$ are learning control vector-valued functions at the $k$th and the $(k+1)$th iterations respectively, $e_{Lk}$, $\dot{e}_{Lk} \in R^m$ are learning error and its derivative vector-valued functions, $\lambda$ is an estimated time advance corresponding to time delay of the process, and $P_L, Q_L \in R^{m \times m}$ are constant learning gain matrices. The discrete version of equation (7.11) is given by

$$v_{k+1}(l) = v_k(l) + P_L \cdot e_{Lk}(l+\mu) + Q_L \cdot c_{Lk}(l+\mu) \tag{7.12}$$

where $l$ is the time index, $\mu$ is an integer representing the advance sampling number, and $c_{Lk}(l)$ is change of learning error defined by $c_{Lk}(l) = e_{Lk}(l+1) - e_{Lk}(l)$.

It is helpful to clarify some points relevant to the above algorithms. First, a variety of special algorithms can be obtained with different selections of learning gain matrices. For example, the simplest version is when $P_L$ is diagonal whereas $Q_L = 0$. In this case, the rate of change of the error is not considered, nor are the loop interacting effects. The learning action in each loop is totally dependent on its own error. It has been demonstrated numerically in Chapter 5 that this algorithm is very effective though simple.

Second, the learning gain matrices $P_L$ and $Q_L$ should be chosen carefully. As would happen in the any other learning system, too large or too small gain values will typically lead to either divergence or an unbearable convergence speed. Thus, a trade-off between speed and convergence should be made. Since we assume that no mathematical model of the process is available, there are no theoretical results available for determining these gains. Fortunately, it has been shown in Chapter 5 that this is not a very hard task because the learning system is relatively robust with respect to the values of the gains.

Third, it is noted that the control update at the $(k+1)$th iteration is based on the information derived at the $k$th iteration, i.e. the previous control action and the resulting performance measured by the error and its derivative. Accordingly, all the information at the $k$th iteration must be stored. Therefore, $e_{Lk}(l+\mu)$ and $c_{Lk}(l+\mu)$ are available for use at the $(k+1)$th iteration, thereby eliminating the necessity for predicting the future states at $l+\mu$. In addition, the other useful by-product from the iteration is that the scaling factors at the $(k+1)$th iteration can be obtained easily by processing the error information stored at the $k$th iteration.

Finally, in the absence of any knowledge about the required control action, it is common practice to assign the initial control $v_0(l)$ to zero at each sampling point. On the other hand, the learning process should be terminated whenever the predefined criterion is satisfied. These stop rules may be an error threshold or a maximum iteration limit.

### 7.3.2 Extracting

Recall that in Section 7.2 we have formulated control rules in the numerical form of $X\_(\hat{u}^j, \delta^j) \rightarrow Y\_(\hat{v}^j, \gamma^j)$. The underlying assumptions about this formulation are fuzzy number interpretation and the intrinsic fuzziness of the BNN. While the effects of $\delta^j$ and $\gamma^j$ can be automatically and implicitly accommodated by the BNN, the primary effort for extracting training rules will be devoted to deriving the central value vectors $\hat{u}^j$ and $\hat{v}^j$.

Although it is possible to use more general and powerful tools to perform this task, for example some traditional clustering algorithms or newly developed neural computations for category learning, here we adopt a much simpler and intuitive scheme. A similar idea has been used in Chapter 5 where two independent rule-bases are required to be constructed.

Suppose that, at the $k_d$th learning iteration, the desired control vector sequence $v_d(l)$ is derived with which the response of the process is satisfied where $l \in [0, I_t]$ and $I_t$ is the maximum sampling number. Meanwhile, the control error $e_c(l)$ is measured, expanded, and denoted by $\bar{u}(l)$ following the controller input formation procedures described in Section 7.2. Combining $\bar{u}(l)$ with $v_d(l)$, a paired and ordered data set $\Gamma_t$: $\{\bar{u}(l), v_d(l)\}$, is created in the STM. Before proceeding, it is worth noting that the formed controller inputs represent a time sequence consisting of a transient stage and a steady-state stage. This observation is very useful in the following discussion.

To obtain a set of rules in the form of equation (7.8) based on recorded data set is basically a matter of converting a time-based $\Gamma_t$ into a value-based set $\Gamma_v$. The number of elements in the extracted set $\Gamma_v$ denoted by $I_v$ should be substantially less than $I_t$+1, the number of elements in the recorded data set. Each data association in $\Gamma_v$ is a representative of several elements in $\Gamma_t$. Accordingly, the extracting process can be thought of as a grouping process during which $I_t$+1 data pairs are appropriately classified as $I_v$ groups, each of which is represented by one and only one data pair. Thus, two functions, grouping and extracting, are involved.

First, each component $\bar{u}_i(l)$ of $\bar{u}(l)$, $l$=1, $I_t$+1, is scaled and quantized to the closed element of the corresponding universe, whereas $v_d(l)$ is left unprocessed for the time being. Thus, a new data set $\Gamma_t^*$: { {u($l$), $v_d(l)$} } is formed, where $u(l) \in U = U_1 \times U_2 \times \cdots \times U_n$ with $U_i$ being a finite and discrete (not necessarily integer) universe, whereas $v_d(l)$ is a continuous-valued vector with each component $v_{dk}$ bounded by $[v_{min,k}, v_{max,k}]$. It is clear that the procedures of scaling and quantizing are, in fact, performing a function of clustering, meaning that several similar $\bar{u}(l)$s may be mapped into one new vector.

Next, the data pairs in the $\Gamma_t^*$ are divided into $I_v$ groups by putting those pairs with the same value of u($l$) into one group, that is,

$$\Gamma_t^* = \Gamma_t^1 \cup \Gamma_t^2 \cup \cdots \cup \Gamma_t^{I_v} \tag{7.13}$$

with

$$\Gamma_t^p : \left\{ \left[ \tilde{u}^p, v_{d1} \right], \left[ \tilde{u}^p, v_{d2} \right] \cdots \left[ \tilde{u}^p, v_{dQ} \right] \right\} \tag{7.14}$$

Note that (1) the time index $l$ in $\Gamma_t^*$ is suppressed since the data are grouped purely by their values; (2) $Q$ different $v_d$ may be associated with the same $\tilde{u}^p$, i.e. $Q \neq 1$, referred to as a conflict group. The reasons for this can be identified as follows. The pure time delays possessed by the controlled process cause the conflict occurring during the transient stage especially when the EC type controller is adopted, whereas the discrete universe $U$ and the continuous universe $V$ are responsible for the conflict occurring during the steady-state stage.

Therefore, the third step is involved in conflict resolution. Here we employ a simple but intuitively reasonable scheme. It is to take the average of $v_{d1}, v_{d2}, \cdots, v_{dQ}$ as their representative, i.e.

$$v = \frac{1}{Q} \sum_{q=1}^{Q} v_{dq} \tag{7.15}$$

Now after conflict resolution, we derive $I_v$ distinct data pairs in a new data set $\Gamma_v$, that is,

$$\Gamma_v = \left\{ \tilde{u}^p, v^p \right\} \qquad p = 1, 2, \cdots, I_v \tag{7.16}$$

By considering that $\tilde{u}^p$ and $v^p$ are the central values corresponding to some fuzzy numbers, the above data pairs can be expressed in the form of "IF $\tilde{u}^p$ THEN $v^p$" which are used to train the BNN-based controller.

## 7.4 Application to multivariable blood pressure control

### 7.4.1 Performance assessment

In order to evaluate the control performances, a set of indices were defined which essentially assess the global performance of the systems. In the first place, the following error vectors are defined:

$$\varepsilon_d = r - y_d \qquad \varepsilon_l = r - y_l \qquad \varepsilon_{fn} = r - y_{fn} \tag{7.17}$$

where $r$ is the command input vector and $y_d$, $y_l$, $y_{fn}$ denote the desired, learned, and fuzzy neural controlled output responses of the process. Then by employing two widely used error integral criteria given

$$\begin{aligned} ISE &= \int_0^\infty \varepsilon^2(t)\mathrm{d}t \\ ITAE &= \int_0^\infty t \cdot |\varepsilon(t)|\,\mathrm{d}t \end{aligned} \tag{7.18}$$

we define the following performance indices:

1. Relative Performance Index (RPI):

$$RPI_i^l = \frac{\phi_{l,i} - \phi_{d,i}}{\phi_{d,i}} \times 100\% \tag{7.19a}$$

2. Normalized Performance Index (NPI)

$$NPI_i^j = \frac{\phi_{fn,i}^j - \phi_{fn,i}^*}{\max_j |\phi_{fn,i}^j - \phi_{fn,i}^*|} \times 100\% \tag{7.19b}$$

3. Locally Normalized Performance (LNP)

$$LNP_i^j = \frac{\phi_{fn,i}^j - \phi_{d,i}}{\max_j |\phi_{fn,i}^j - \phi_{d,i}|} \times 100\% \tag{7.19c}$$

where $\phi_i$, $i = 1, 2, 3, 4$, are error integrals in the order $\{\phi_1, \phi_2, \phi_3, \phi_4\}$ corresponding to {ISE_CO, ITAE_CO, ISE_MAP, ITAE_MAP}, and superscripts $j$ and * stand for cases which will be explained shortly.

Basically $RPI_i$ are used to assess the learning ability during the learning stage; $NPI_i$ are employed to investigate the generalizing capacity of the BNN; and $LNP_i$ are used to compare the performances of the three types of controllers.

### 7.4.2 Simulation results

The main aim of the simulations is to examine the behaviours of the proposed scheme in learning, extracting, and operational stages when applied to the problem of multivariable blood pressure control as described in Appendix II. Because the learning algorithms have been treated extensively in Chapter 4, here particular attention is placed on the investigation of the property of robustness possessed by the BNN-based controller which is built on the basis of the learned rules. In addition, comparative studies of three types of controllers were carried out, again with a viewpoint of investigating performance robustness.

Throughout this work, the following values were used. The learning interval contains 101 sampling points with sampling time being 30 s. Set-points for $CO$ and $MAP$ were set to be 20 ml/s and -10 mmHg changing from normal values of 100 ml/s and 120 mmHg respectively. The learning matrix in equation (7.12) was diagonal with $P_L = \text{diag}\{0.06, -0.07)$ and $Q_L = 0$. The number of iterations during the learning stage were 15. In addition, 15 hidden units were used for all the controllers.

The first set of simulations was conducted to inspect the performance of the systems during the learning and extracting stages. Table 7.1 summarizes the results. By varying the parameters $\xi$, $t_s$ in the second-order reference models (7.10) with $t_s$ being the settling time approximately given by $t_s \approx 4/\xi w_n$, the learning ability was assessed via the performance indices $RPI_i$. In most cases, the relative differences were less than 10% and even became negative, indicating that the learning teacher is reliable in following the performance requirements specified by the reference models. Corresponding to these four cases, a set of rules was constructed as shown in Table 7.1, each of which was extracted from 101 data pairs and related to EC, ES, and ECS type controllers respectively. As expected, the rule numbers of ES and ECS are slightly more than that of EC due to the property of the step command input.

The second group of simulations was concerned with examining the robust capability of the trained controller with respect to variations of the process parameters during the operational stage. This problem may be regarded as equivalent to investigating the generalization property of the trained network. By selecting the EC type controller as a test bed with the performance requirements corresponding to the third case in Table 7.1, a controller was realized after learning, extracting, and training stages. Initially, with the same process parameters as used during the learning phrase, a set of error integrals was derived and denoted as $\phi^*_{fn,i}$ by operating the trained EC controller. Then the process parameters were changed from their normal values and the weight-frozen controller was applied again with this altered environment. Thus, a set of new error integrals was obtained and denoted as $\phi^j_{fn,i}$, where $i = 1, 2, 3, 4$ and $j = 1, 2, \cdots, 9$ representing nine different cases as indicated in the first column in Table 7.2.

The resultant performance indices $NPI_i^j$ normalized within each column, together with averaged values $\overline{NPI}$, with respect to nine cases are listed in Table 7.2. In the

**Table 7.1** Learning performance and rule number

| $\xi_{co}$ | $\xi_{map}$ | $t_{co}$ | $t_{map}$ | $RPI_1$ | $RPI_2$ | $RPI_3$ | $RPI_4$ | EC | ES | ECS |
|---|---|---|---|---|---|---|---|---|---|---|
| 0.8 | 0.8 | 240 | 240 | 9.76 | 21.56 | -6.28 | 5.67 | 9 | 9 | 11 |
| 0.9 | 0.9 | 300 | 300 | 5.20 | 8.10 | -3.90 | 5.10 | 13 | 13 | 15 |
| 0.9 | 0.7 | 300 | 240 | 2.00 | 7.70 | 1.69 | 14.1 | 11 | 12 | 13 |
| 0.7 | 0.9 | 240 | 300 | 16.56 | 37.70 | -11.30 | 1.8 | 10 | 12 | 12 |
| Averaging | Value | | | 8.00 | 18.76 | -4.94 | 6.70 | 11 | 12 | 13 |

table the first six cases are related to variations of time constants of up to 20% change in values, whereas the last three cases were involved with variations of time delays of up to 30 s (one sampling instant) change. It can be seen that the controller is less sensitive to the variations of the time constant than to those of time delays. One may think that, due to the utilization of the normalized measures $NPI$, this table has not indicated how poor the actual responses might be. To clarify this point, the output responses corresponding to the worst case in Table 7.2 are depicted in Figure 7.3. It appears that, regardless of the large averaged performance variation (90.61%), the responses are actually very acceptable in the sense of closeness to the desired responses.

The last set of simulations focus on the issue of comparisons between the three types of controllers with different input variables. The comparative studies were made on the basis of performance robustness with respect to variation of process gains and the effect of a noisy environment. By fixing the requirement performances to be the same for all three types of controller, a set of results was obtained and is shown in Table 7.3, where the performance indices $LNP_i$ were used, and the process gains were increased by 5% from their nominal values. Corresponding to EC, ES, and ECS controller types, each entry in the table is composed of three values which are normalized locally for the purpose of comparison. The last three columns are the averaged values with $\overline{LNP}_{1,3}$ and $\overline{LNP}_{2,4}$ being associated with $ISE$ and $ITAE$ respectively and $\overline{LNP}$ being the average of $\overline{LNP}_{1,3}$ and $\overline{LNP}_{2,4}$. The output responses corresponding to the nominal case in Table 7.3 and both output responses and control signals in a noisy environment are shown in Figures 7.4, 7.5, 7.6, and 7.7 respectively. The following conclusions can be drawn from the table and figures:

**Table 7.2** Performance robustness of EC controller type

| CASE | $NPI_1$ | $NPI_2$ | $NPI_3$ | $NPI_4$ | $\overline{NPI}$ |
|---|---|---|---|---|---|
| $T_{c1} = 100.92 \quad T_{c2} = 58.75$ | 12.0 | -0.80 | -4.95 | 23.95 | 7.55 |
| $T_{c1} = 84.10 \quad T_{c2} = 70.50$ | 9.60 | 3.80 | 57.00 | 42.00 | 28.10 |
| $T_{c1} = 100.92 \quad T_{c2} = 70.50$ | 23.00 | 3.90 | 27.20 | 21.20 | 18.83 |
| $T_{c1} = 67.28 \quad T_{c2} = 47.00$ | -16.50 | 1.80 | -11.30 | 23.90 | 5.13 |
| $T_{c1} = 100.92 \quad T_{c2} = 47.00$ | 8.60 | 0.40 | -19.00 | 16.70 | 6.70 |
| $T_{c1} = 67.28 \quad T_{c2} = 47.00$ | -4.60 | 3.40 | 100.00 | 57.70 | 39.12 |
| $\tau_1 = 90 \quad \tau_2 = 30$ | 96.00 | 100.00 | 84.00 | 82.46 | 90.61* |
| $\tau_1 = 90 \quad \tau_2 = 60$ | 18.00 | 63.00 | 95.00 | 100 | 69.00 |
| $\tau_1 = 60 \quad \tau_2 = 60$ | 100.00 | 6.70 | 83.00 | 32.00 | 55.42 |

*The output responses corresponding to this worst case are shown in Figure 7.3

1. For the case of nominal process gains, the transient responses of the EC controller type seem to be the best, indicated by the smallest $\overline{LNP}_{1,3}$. This can also be seen by inspecting the actual responses as illustrated in Figure 7.4, where the initial responses are fast but are suppressed while approaching the set-points, leading to irregular response shapes. The result is conceptually related to the derivative term being most sensitive to fast-changing signals.
2. Again for the case of nominal process gains, ES and ECS controller types have better steady-state responses than EC. This is clearly perceived by the values of $\overline{LNP}_{2,4}$ and Figure 7.3. The conclusion is consistent with the conceptual consideration that an integral term (sum in this case) is an effective way of eliminating the steady-state errors.

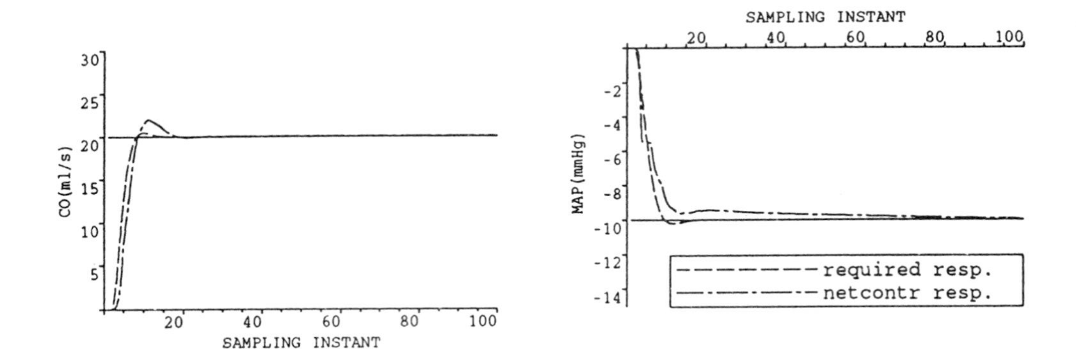


**Figure 7.3** Outputs of the process corresponding to the worst case in Table 7.2

**Table 7.3** Performance comparison of EC, ES, and ECS controller types

| CASE | TYPE | $LNP_1$ | $LNP_2$ | $LNP_3$ | $LNP_4$ | $\overline{LNP}_{1,3}$ | $\overline{LNP}_{2,4}$ | $\overline{LNP}$ |
|---|---|---|---|---|---|---|---|---|
| NOMINAL VALUES | EC<br>ES<br>ECS | 33.9<br>100.0<br>96.0 | 99.0<br>100.0<br>12.8 | -100.0<br>-84.0<br>-86.5 | 100.0<br>-2.8<br>-4.4 | -33.0<br>8.0<br>4.7 | 99.5<br>48.6<br>4.2 | 33.3<br>28.3<br>4.5 |
| $\Delta K_{11}$ | EC<br>ES<br>ECS | 27.0<br>100.0<br>97.0 | 62.6<br>95.5<br>100.0 | 100.0<br>-31.3<br>-37.5 | 100.0<br>22.0<br>12.0 | 63.0<br>35.0<br>29.7 | 80.0<br>58.75<br>56.0 | 71.5<br>46.9<br>42.9 |
| $\Delta K_{12}$ | EC<br>ES<br>ECS | 29.0<br>100.0<br>97.0 | 44.7<br>90.9<br>100 | 40.8<br>-91.0<br>-100 | 100<br>22.0<br>5.0 | 34.9<br>4.5<br>9.5 | 72.0<br>56.4<br>52.5 | 53.5<br>30.6<br>31.0 |
| $\Delta K_{21}$ | EC<br>ES<br>ECS | 26.9<br>100.0<br>94.0 | 13.5<br>100.0<br>69.0 | 29.0<br>-88.0<br>-100.0 | 100.0<br>5.7<br>8.0 | 27.95<br>6.0<br>3.0 | 56.78<br>52.8<br>38.5 | 42.4<br>29.4<br>20.8 |
| $\Delta K_{22}$ | EC<br>ES<br>ECS | 30.0<br>100.0<br>93.0 | 49.0<br>100.0<br>82.9 | 100.0<br>-6.0<br>-6.0 | 100.0<br>3.8<br>2.9 | 65.0<br>47.0<br>43.5 | 74.5<br>51.9<br>42.9 | 69.5<br>49.5<br>43.2 |

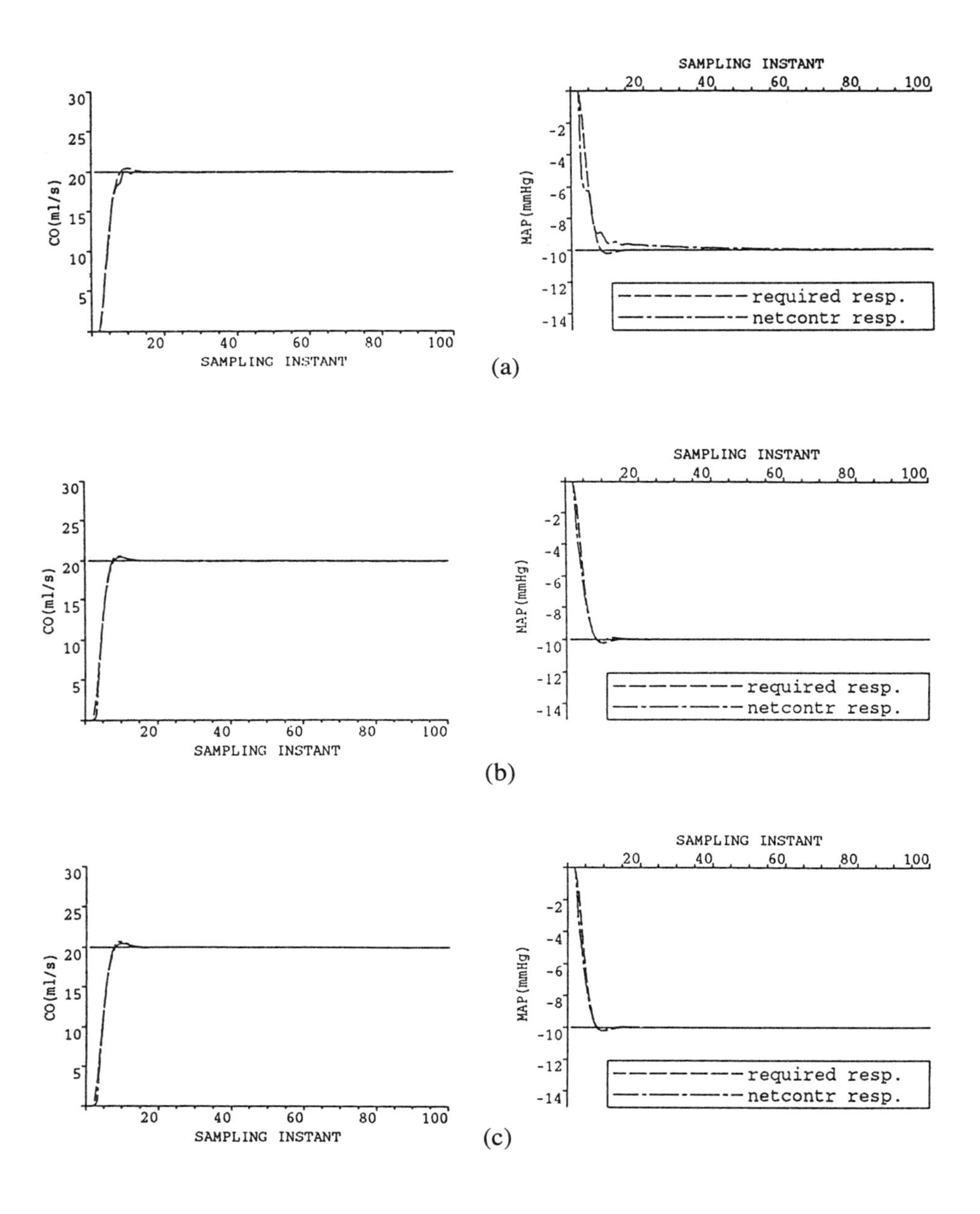


**Figure 7.4** Outputs of the process corresponding to the nominal case in Table 7.3: (a) EC type; (b) ES type; (c) ECS type

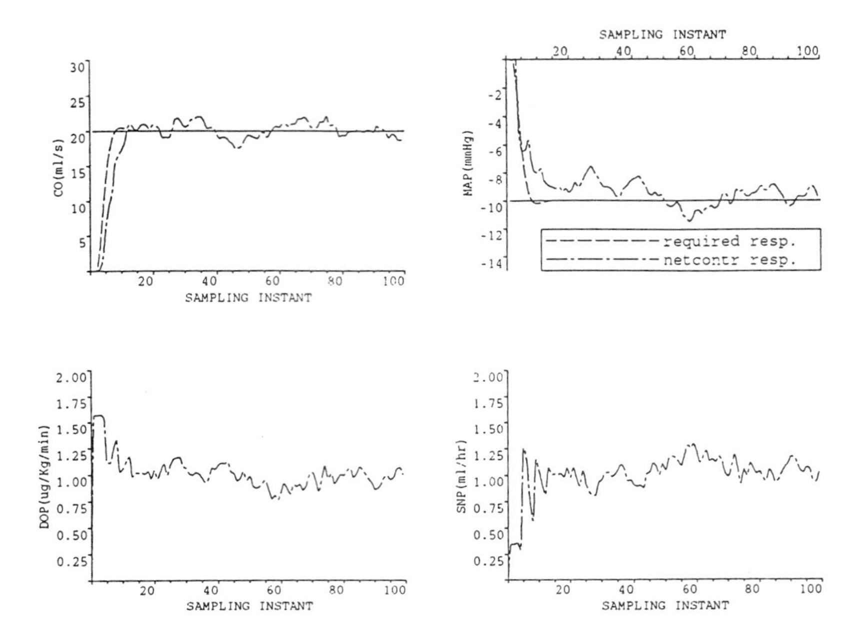


**Figure 7.5** Outputs and inputs of the process in a noisy environment: EC type

3. By looking at the last column of Table 7.3, it is easy to see that the ECS controller type is the most robust in regard to variations of gains, as indicated by the defined indices. In contrast, as would be expected, the EC controller type is most sensitive.
4. The outputs of the ES controller are the most smooth, ECS outputs are less smooth, and the least smooth signals occur in the EC controller. This is demonstrated clearly in Figures 7.5, 7.6, and 7.7 where the measured process outputs were assumed to be noise-contaminated.
5. The above comments seem to suggest that the ECS type is the best candidate because of its good robustness and smooth control, which are desired properties in practice. However, the ES type may be acceptable also because it is less complicated than ECS especially when multivariable processes are involved.

## 7.5 Discussion

It is worthwhile discussing some important and interesting points implied in the proposed method. Here we wish to mention only two of them relating the present scheme to more traditional ones. The first point relevant to the existing fuzzy

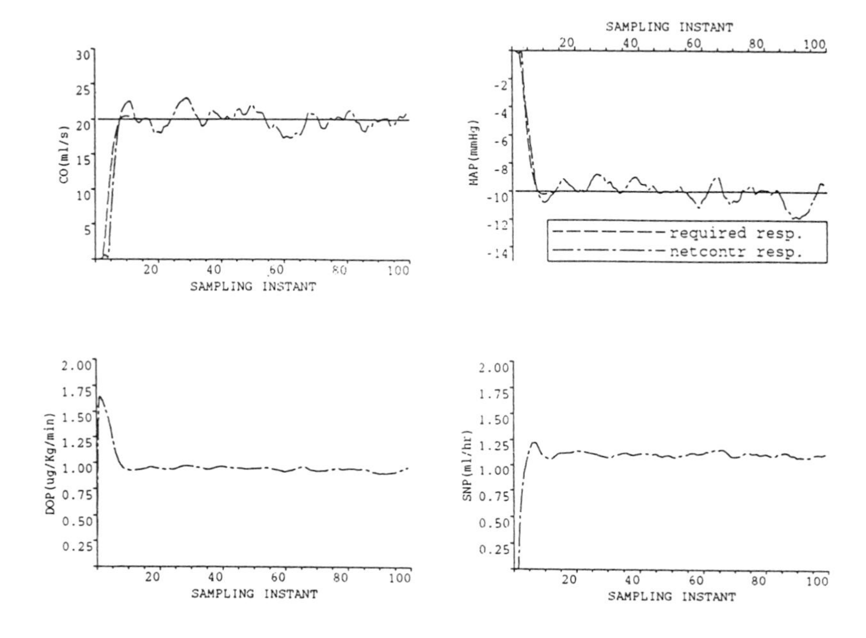


**Figure 7.6** Outputs and inputs of the process in a noisy environment: ES type

controllers concerns the problem of the acquisition of the control rules, whereas the second associated with traditional PID controllers involves the controller's computational aspect.

From the viewpoint of implementation, the proposed method provides a much more simple, flexible and general mechanism for extracting control rules than the existing fuzzy controllers. In principle, our method can deal with the situation of processes with an arbitrary number of variables and an associated controller with an arbitrary number of input variables. In contrast, the majority of existing applications of fuzzy controllers only involved single-input single-output (SISO) problems although some authors have advocated that the generalization to MIMO systems is straightforward, a statement, which we think is not true in general. Furthermore, it is interesting to note that existing fuzzy controllers have almost always been dominated by the EC type. The above two phenomena arise not from the lack of an efficient fuzzy controller algorithm, but from the difficulty in deriving the required rules.

While a skilled control engineer may be able to summarize the control rules for a SISO process according to, for instance, the states of error and error-in-change (Yamazaki and Sugeno 1985), it is far beyond the imagination of any control engineer to articulate a centralized ECS control rule-base (e.g. with six inputs and two outputs) for a multivariable process with strong interaction between the

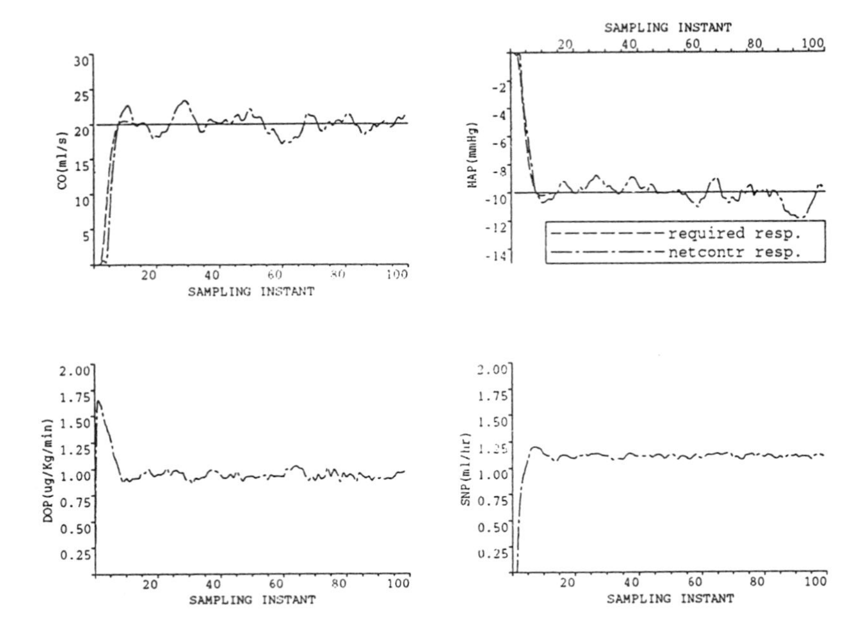


**Figure 7.7** Outputs and inputs of the process in a noisy environment: ECS type

variables. Likewise, the existing self-organizing schemes are not general enough to handle the problem posed above. For example, with an assumption of availability of a crude process model, the method suggested by Procyk and Mamdani (1979) may be feasible for building the rule-base for a multivariable process with a set of decentralized controllers for each loop and each controller being of the EC type. However, it is extremely difficult, if not impossible, to construct a rule-base for ECS type controllers even for each independent loop because of the difficulty of derivation of a ECS type performance index table (PIT) which is the basis of this method. In contrast, the simulation study reported in the previous section indicates that the proposed scheme can deal with the problem without much effort, suggesting that this method is superior to human experts and to the existing methods in the sense discussed above.

Now we turn our attention to the second point. It has been known that fuzzy controllers generally produce a comparable or even better control performance than traditional PID controllers. Conceptually, this is because a fuzzy controller usually performs a nonlinear mapping from its inputs to its outputs, whereas a fixed-gain PID controller is linear. Because our fuzzy controller is built on the structure of layered feedforward networks, it can naturally carry out an arbitrary nonlinear mapping. In this respect, the BNN-based ECS fuzzy controller, having similar inputs to the PID controller, definitely possesses a much more powerful functional mapping capacity than the traditional PID controller, thereby enabling the former to handle more complicated and nonlinear control problems provided, of course, that the training rules are available.

## 7.6 Summary

By making a crucial but realistic assumption that neither control experts nor teacher signals are available, we have presented a novel and systematic design approach which is capable of learning and extracting required control rules automatically from the controlled environment for use with BNN-based fuzzy controllers. Three possible controller structures have been suggested and compared via simulations. We have made some remarks relating the present method to other traditional ones. The simulation results of blood pressure control have demonstrated the utility and feasibility of the proposed approach in solving relatively complex control problems, in particular, those problems where neither control experts nor mathematical models of the controlled process are available.

CHAPTER 8

# A hybrid neural network-based-self-organizing fuzzy controller

*A hybrid fuzzy control system with abilities for self-organizing and self-learning is developed, and consists of a variable-structure competitive network (VSC) and a standard BNN network.*

## 8.1 Introduction

It has been demonstrated in Chapter 6 that approximate reasoning in rule-based fuzzy control systems can be implemented by BNN neural networks. We have also taken one step further in Chapter 7 by assuming that no experts are available for training the BNN-based fuzzy controller, and proposed a general mechanism capable of extracting training rules automatically from the process being controlled by incorporating a simple learning algorithm into the system. The approach consists of four stages, namely, learning, extracting, training, and operation. While the feasibility of this approach has been verified numerically, it is still unsatisfactory because of the multi-stage requirement.

In an attempt to develop a multivariable controller capable of self-organizing suitable rule-bases with the little prior knowledge about the process, this chapter proposes a controller departing significantly from the previous ones mainly in the following ways. First, instead of completing the control task by using several stages, we carry out the rule-base construction in an on-line control manner without involving any intermittent procedures. Second, we look at the network-based controller not only from the computational viewpoint, but also from knowledge representation and processing perspectives. Therefore, while taking advantage of the computational efficiency of the network paradigm, we maintain the clarity for the rule-base paradigm, thus providing an explicit explanation facility of the network action.

The proposed self-organizing and self-learning multivariable fuzzy control system has an architecture similar to that used by traditional model reference adaptive control systems (MRAC). The controller is built as a hybrid neural network consisting of a variable-structure competitive network and a standard BNN network. Instead of using tapped delay signals as the controller inputs, as almost exclusively adopted in neurocontrollers, a widely-used PID-like feedback controller structure is employed here. The required self-learning function is carried out by a modified back-propagation algorithm similar to the one proposed by Psaltis *et al.* (1988), whereas the self-organizing function is accomplished by cascading a variable-structure competitive network (VSC) with a BNN network. The overall system can be operated in a real-time manner under a closed-loop and multivariable environment.

The rest of the chapter is organized as follows. In Section 8.2 the system structure is presented briefly. The underlying self-organizing principles are described in Section 8.3. Section 8.4 is dedicated to the development of an extended back-propagation algorithm functioning self-learning connection weights in the BNN. The simulation results are given in Section 8.5.

## 8.2 System structure

Figure 8.1 shows a schematic of the proposed system structure. It consists of a basic feedback control loop containing a hybrid network-based fuzzy controller, a controlled process, and a performance loop composed of reference models and learning laws. Assuming that neither a control expert nor a mathematical model of the process is available, the objectives of the overall system are (a) to minimize the tracking error between the desired output specified by the reference model and the actual output of the process in the whole time interval of interest by adjusting the connection weights of the networks, and meanwhile (b) to construct control rule-bases dynamically, by observing, recording, and processing the input and output data associated with the net-controller. It should be noted that the above objectives are achieved only gradually with increasing trial. Furthermore, the whole system performs the two functions of control and learning simultaneously. Within each sampling period, a feedforward pass produces the suitable control input to the process and the backward step learns both control rules and connection weights. Thus, it is very important to note that the operation of the system is along both time and space dimensions. The function of each component in Figure 8.1 is described below.

*Process.* We will make the same assumptions about the process being controlled as made in Chapter 7.

*Input formation.* It is assumed that the input to the controller can be EC, ES, or ECS types as defined in Chapter 7. In what follows, we will denote the input by $u$ which is assumed to be an $n$-dimensional vector.

*Reference model.* The desired transient and state-steady performances are signified by the reference model which provides a prototype for use by the learning mechanism. Although the design of reference models for multivariable systems can

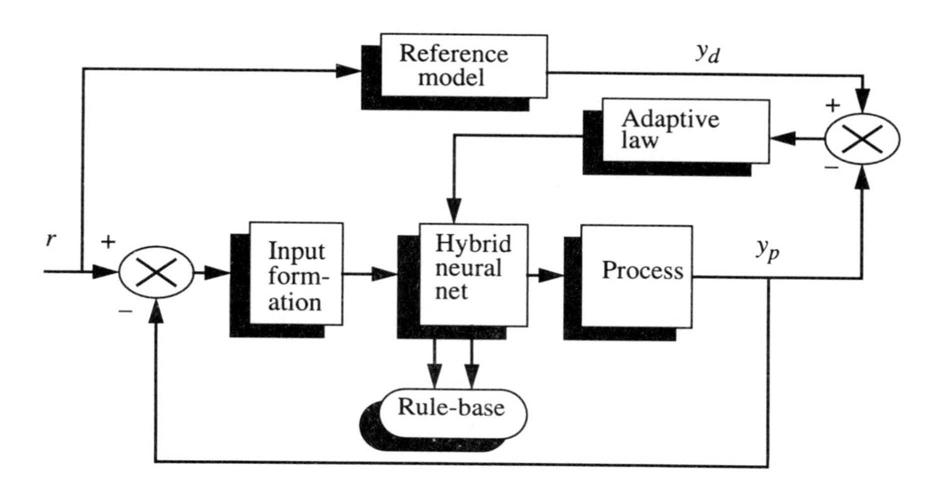


**Figure 8.1** Overall system structure

be complicated, here again for the sake of simplicity we adopt the reference model whose transfer matrix relating the command signal $r$ to the model output $y_d$ is diagonal with the elements being the second-order linear models. It should be emphasized, however, that the noninteracting and linear assumptions in the model do not require that the process itself must have these properties. Yet they do suggest that some knowledge about the process should be available. For example, we implicitly assume that each process output is primarily affected by one of the $m$ inputs so that $m$ inputs–outputs are paired in the model.

*Hybrid neural networks.* As a key component of the system, the hybrid neural network, as shown in Figure 8.2, is designed to stabilize the dynamical input vector $u$, compute the present control value, and provide the necessary information for the rule-base formation module. To be more specific, the VSC network converts a time-based on-line incoming sequence into value-based pattern vectors which are then broadcast to the BNN network, where the present control is calculated by the BNN using feedforward algorithms. Thus, the VSC can be regarded as a preprocessor or a pattern supplier for the BNN.

From the knowledge processing point of view, the networks not only perform a self-organizing function by which the required knowledge bases are constructed automatically, but also behave as a real-time fuzzy reasoner in the sense that the present control is deduced by processing the encoded and distributed knowledge in the network with the current input. It has been claimed that the BNN network inherently possesses some fuzziness due mainly to its distributiveness, whereby the function of approximate reasoning can be performed effectively by the feedforward

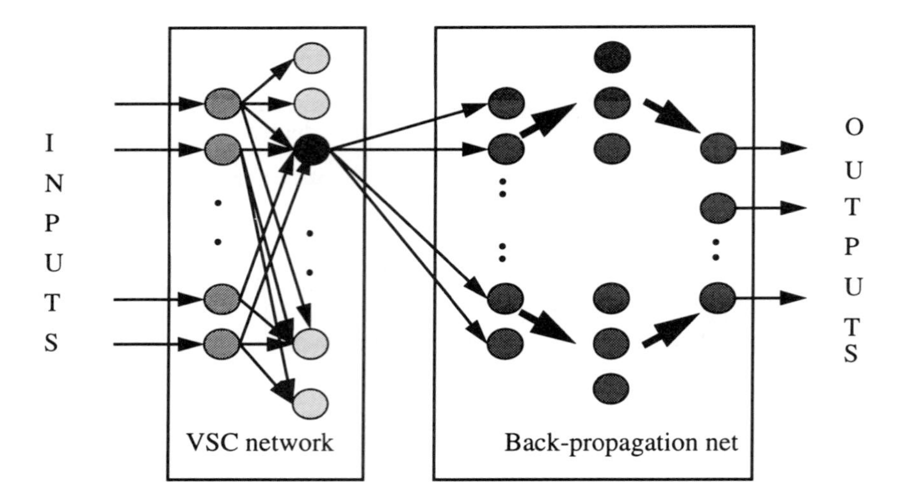


**Figure 8.2** Hybrid neural networks

manipulation of the BNN. It is in this functional sense that we say the fuzzy controller is equivalent to the BNN network.

*Rule-base formation.* By appropriately recording the input and output of the hybrid neural networks, the formed rule-base, each rule having the format of IF *situation* THEN *action*, offers an explicit explanation of what the net-based controller has done. This can be understood in the following way. Because the information used to guide the adjustment of the connection weights of the BNN is similar to that for constructing the IF-THEN rules, the mapping property of the resultant BNN must exhibit similar behaviour to that of the IF-THEN conditional statements. More specifically, when the IF part of a rule is presented to the BNN, it should produce an output which is similar to the THEN part of that rule. Again, from the knowledge representation viewpoint, both paradigms in effect represent the same body of knowledge using the different forms. The BNN represents it distributedly and implicitly by the connection weights, whereas the rules represent the knowledge concisely and explicitly by relating the situation to the corresponding action.

*Adaptive mechanism.* While the VSC network can be self-organized by merely relying on the incoming information, the weights of the BNN network must be adjusted by back-propagating the predefined learning errors. Due to the unavailability of the teacher signals, a modified and simple adaptive law has been developed and will be detailed in Section 8.4.

## 8.3 Dynamical self-organizing

This section will focus on the VSC networks with emphasis placed on its dynamical self-organizing process. Before proceeding, the traditional competitive network is described next.

### 8.3.1 Competitive learning neural networks

Competitive or unsupervised learning systems have been a very rich research topic (e.g. Grossberg 1987; Rumelhart and Zipers 1985; Kohonen 1990; Kosko 1991; Lemmon 1991) It can be used as a mechanism for adaptive (learning) vector quantization (AVQ or LVQ) in which the system adaptively quantizes the pattern space by discovering a set of representative prototypes. Therefore, the trained system can carry out the task of classification or recognition (e.g. Kohonen 1988b). Viewing the system as performing an estimation of the centroids or modes of the sampled probability density function, the scheme may also serve as a basis for developing a kind of parameter estimation algorithm (Lemmon 1991)

There exist a number of variations of AVQ with different net configurations or algorithms. However, the basic idea is quite similar, that is, to categorize vectorial stochastic data into different groups by employing some metric measures with a winner-selection criterion. An AVQ network is usually composed of two fully connected layers, an input layer with $n$ units and a competitive layer with $p$ units. The input layer can be simply receiving the incoming input vector $u(t) \in R^n$ and forwarding it to the competitive layer through the connecting weight vectors $w_j(t) \in R^n$. The competition takes place in the following way. At each time, the input $u(t)$ is compared with weight vectors $w_j$ according to some predefined metric criterion. If $w_J$ associated with the $J$th unit in the competitive layer is closest to the current $u(t)$ in the defined metric sense, the $J$th unit is regarded as the winner. Following the competition, in accordance with a law of "winner-take-all" the activation value $I_j$ of all units are set to be 0 except the winner unit $J$ whose activation values $I_J$ is set to be 1. The winner's weight is updated by adding a scaled difference $u(t) - w_J(t)$ to $w_J(t)$ such that the modified $w_J(t+1)$ is closer to $u$ and therefore, more representative of $u$, whereas the modification of the weights of the remaining units is laterally inhabited.

Assume that the number of units, $M$, at the competition layer is predetermined according to the nature of the task. For example, $M$ can be equal to the number of pattern-classes in the sample space. In addition, assume that the training samples $u(t)$ are drawn randomly in accordance with a fixed probability distribution. Then, the algorithm is mathematically presented below where the simpliest update algorithm is adopted. For other more complicated algorithms, we refer readers to Kohonen (1990) and Kosko (1991).

*Competitive algorithm:*

1. Initialize weight vectors $w_j(0)$. There exist some schemes concerning this matter (Kohonen 1990). For instance, $w_j(0)$ can be set to be equal to the first $M$ samples, that is, $w_j(0) = u(j)$, $j = 1, 2, \cdots, M$.
2. Calculate the distance metric of the current incoming sample with respect to $M$ units using the formula $\| w_j(t) - u(t) \|$, where $\| \,.\, \|$ denotes a norm which can be defined in a number of ways.
3. Determine the winner among $M$ units by finding the unit closest to the $u(t)$ in the sense of having a minimum distance:

$$\|w_J(t) - u(t)\| = \min_j \|w_j(t) - u(t)\| \tag{8.1}$$

4. Update the weight vectors $w_j(t)$ by

$$\begin{cases} w_j(t) = w_j(t-1) + \alpha_t \cdot [u(t) - w_j(t-1)] & \text{if} \quad j = J \\ w_j(t) = w_j(t-1) & \text{if} \quad j \neq J \end{cases} \tag{8.2}$$

where $0 < \alpha_t < 1$ is a gain sequence decreasing monotonically with time in a linear or exponential manner. It is well known (e.g. Duda and Hart 1973) that, in the context of stochastic approximation, if the coefficients $\alpha_t$ satisfy

$$\lim_{p \to \infty} \sum_{t=1}^{p} \alpha_t = +\infty \tag{8.3}$$

$$\lim_{p \to \infty} \sum_{t=1}^{p} \alpha_t^2 < \infty \tag{8.4}$$

then convergence to local minima can be guaranteed in the sense of mean squared performance measure. The requirements of equations (8.3) and (8.4) imply that the sequence $\alpha_t$ should tend to 0 at an appropriate speed, i.e not too fast. Similarly, Clark and Ravishankar (1990) have proved recently that algorithm (8.2) can be assured to converge in probability to the probabilistic centroid of the training sample set if and only if $\alpha_t$ converges to 0 and series $\sum \alpha_t$ diverges. In this study, a harmonic series $\alpha_t = 1 / t$ satisfying the above mentioned conditions will be used.

*Remark*: the competitive algorithm presented above is used for discovering $M$ prototypes from a data set. It can be roughly regarded as a process of clustering the data into $M$ groups. Compared with the other well-known performance-based clustering schemes such as the c-means algorithm, the above algorithm has the merits of simplicity and easy implementation. As pointed out by Kohonen (1990), the learning algorithm (8.2) in fact can be considered as minimizing an average squared discretization error function.

### 8.3.2 Variable-structure competitive (VSC) algorithm

The algorithm outlined above is basically an off-line procedure and cannot be applied directly to our case where on-line operation is essential. This infeasibility stems primarily from the fact that the number of cluster centroids, or equivalently control rules in this context, is generally either unknown in advance or not unique. Thus, a problem arises as to how the input space should be partitioned reasonably and effectively. In the case of having a low-dimensional input space, say $R^n$ with n $\leq 2$, it may be possible to predetermine the number of subspaces by taking the property of the input data into account and defining the input space as being finite and perhaps discrete. However, the approach turns out to be inadequate for dealing with multivariable control problems with a high-dimensional input space. For example, there will be $n = 6$ inputs for a $2 \times 2$ controlled process with ECS controller type. In addition, the distribution feature of the input data varies from case to case depending largely on the characteristics of the controlled process, the form of command inputs and performance requirements. In general, the input data are sparsely distributed over the whole input space. Thus, higher dimensionality of the input space together with uncertainly and sparsely distributed properties of the on-line sampled data make the above mentioned method not only difficult but also undesirable to apply.

Consequently, it is necessary to develop a learning system capable of self-organizing both class numbers and associated representative vector values. The algorithm to be presented below is an attempt to meet this objective. Starting from the empty units at the competitive layer, the input space is automatically partitioned and adaptively quantized in response to the on-line incoming data. This process proceeds continuously with increasing iteration, and therefore, the number $M$ of the units at the competitive layer grows accordingly. A key idea is to treat each subspace as a completely localized region by introducing a valid radius and a local gain with respect to each subspace. More specifically, each unit denoted by $\Phi_j$ is characterized by four parameters: $\Phi_j = \{w_j, \Delta_j, \alpha_j, v_j\}$, where $w_j$ denotes connecting weight vectors from input $u$ to the $j$th unit $\Phi_j$, $\Delta_j$ is a valid radius defining a neighbourhood for $\Phi_j$ and can be specified in advance or adjusted adaptively like that of $w_j$, $\alpha_j$ is the local gain controlling the speed of adaptive process of $w_j$ and is inversely proportional to the active frequency $n_j$ of unit $\Phi_j$ up to the present time instant, and $v_j$ is the current controller output vector from the BNN net in response to $w_j$ and regarded as the THEN part of the $j$th control rule. It should be pointed out that the output vectors of the BNN at some successive sampling instants can be different in response to the same $w_j$. In this case $v_j$ should be the average of these outputs corresponding to the same $w_j$.

In view of the above discussion, the operating procedures of the VSC competitive algorithm can be described as follows. With each iteration and each sampling time, the processed control error vector $u$ is received. Then a competitive process takes place among the existing units in regard to the current $u$ by using the winner-selection criterion. If one of the existing units wins the competition, the corresponding $w_j$ is modified; otherwise, a new unit is created with an appropriate initialization scheme. Finally, the modified or initialized weight vector is broadcast to the input layer of BNN net. By assuming that all $\Delta_j$ are identical, i.e $\Delta_j = \delta$, the proposed algorithm is presented formally below, where $k$, $t$, and $T$ denote the iteration number, the sampling instant, and maximum sampling time respectively.

*Variable-structure competitive algorithm:*

1. Initialization. $w_1^1(0) = u(0)$; $n_1^1(0) = 1$; $\alpha_1^1(0) = 1/n_1^1(0)$; $M^1(0) = 1$.
2. At the $k$th iteration, do the following at each sampling time $t$:
   (a) Find the unit $J$ which has the minimum distance to the current $u$ by

$$D(w_J^k,u) = ||w_J^k(t) - u(t)|| = \min_{j=1,\,M} ||w_j^k(t) - u(t)|| \tag{8.5}$$

   (b) Determine the winner using the following rule:

$$\begin{cases} \text{if} \quad D(w_J^k,u) \le \delta & \rightarrow \textit{J is the winner} \\ \text{if} \quad D(w_J^k,u) > \delta & \rightarrow \textit{create a new unit} \end{cases} \tag{8.6}$$

   (c) Modify or intialize parameters:
   If $J$ is the winner:

$$\begin{cases} n_J^k(t) = n_J^k(t-1) + 1; \quad \alpha_J^k(t) = \dfrac{1}{n_J^k(t)} \\ w_J^k(t) = w_J^k(t-1) + \alpha_J^k \cdot [u(t) - w_J^k(t-1)] \\ M^k(t) = M^k(t-1) \end{cases} \tag{8.7}$$

   If a new unit is created:

$$\begin{cases} M^k(t) = M^k(t-1) + 1 \\ w_{M^k(t)}(t) = u(t) \\ n_{M^k(t)}^k(t) = 1 \end{cases} \tag{8.8}$$

   (d) Output $w_J^k$ or $w_{M^k(t)}^k$ to the BNN input layer.
3. After the $k$th iteration, remove all inactive units and reinitialize all active units:

$$\begin{cases} M^{k+1}(0) = M^k(T) - M_{inactive}^k \\ w_j^{k+1}(0) = w_j^k(T) \\ \alpha_j^{k+1}(0) = \alpha_j^k(T) \\ j = 1, 2, \cdots, M^{k+1}(0) \end{cases} \tag{8.9}$$

It should be noted that step 3 is a check procedure which is intended to retain only those units which have been active at least once during the latest iteration, and to discard all other inactive units. This step is important not only for keeping a reasonable net size, but also for maintaining and eventually, after the iteration has terminated, deriving a satisfactory rule-base in which each rule makes its own contribution to fulfil present performance requirements.

*Remark 1:* We note that the input space is partitioned in a soft or fuzzy manner. Defined by a valid radius, each cluster centroid occupies a hyper-region, part of which is shared by other existing neighbouring clusters. This reflects a sharp distinction to that of the cluster number-fixed competitive process where the input space is disjointly partitioned and each point in the space belongs to its nearest cluster.

*Remark 2:* The partition of the input space is a dynamical process featuring the variable-structure competitive property. This can be accounted for by the fact that not only the value of each representative vector is dynamically modified, but also the number of the vectors is dynamically adjusted with time. The variable property is further enforced along the iteration dimension as indicated by the last step in the above algorithm. Thus, the evolving process of the network involves both temporal and spatial aspects.

*Remark 3:* After the iteration is terminated with satisfactory control performance, the units at the competitive layer can be considered to represent rules by taking one-to-one correspondence between a set of associated parameters $\{w_j, \delta, v_j\}$ and a rule in the form of "IF $(w_j, \delta)$ THEN $v_j$". As discussed previously, the constructed rule-base paradigm provides an explicit explanation for the behaviour of the BNN since the principal function which the distributed BNN can perform is summarized by a centralized rule-base. In addition, the rule-base may be used as a basis for algorithm-based fuzzy controllers or even as a look-up table directly for the current control problem provided that the valid radius is sufficiently small.

## 8.4 Adaptive mechanism

As is well known, the standard back-propagation training algorithm is designed to minimize a predefined error norm concerning all the training patterns by employing a gradient descent procedure. The error is defined to be the difference between the desired and actual output of the network. This implies that a teacher signal at the output layer must be supplied for performing the task. Unfortunately, this required signal is generally unavailable when the BNN is used as a controller as configured in Figure 8.1. On the contrary, every effort should be made to design a controller (neural-like or not) to generate this signal so as to drive the process output to satisfy the predefined performance requirements. In the present context, the problem becomes how to modify the weights of the BNN so that the process output $y_p$ is close enough to the desired output $y_d$ specified by the reference model in some meaningful sense. Therefore, instead of back-propagating the error at the output of the BNN, we back-propagate the error at the output of the process, thereby providing adaptive signals for adjusting the BNN parameters.

However, the problem seems to remain unsolved in the sense of how the error can be related or back-propagated to the BNN adjustable weights. This is a problem of credit assignment, i.e. a problem of how the reward (blame) is distributed to individual weights contributing to the present error. Intuitively, the solution must rely on some knowledge about the controlled process. Indeed one way to overcome this difficulty is to first identify the model of the process using a BNN structure identical to that of the controller, then by the *certainty equivalence principle*, assuming that the model is a true representation of the process, the tracking error can be back-propagated readily through this weight-fixed network layer by layer down to the controller. Realizing that this approach consisting of two separate stages is complex and a slow design process, there do exist other possibilities for handling this matter. For example, Psaltis *et al.* (1988) have suggested a method in which the tracking error is directly back-propagated through the process itself with an assumption that the forward Jacobian of the process is known. By adopting a similar idea, we argue, however, that the required Jacobian matrix can be replaced by an extremely simple approximation form, a diagonal matrix, thereby keeping the knowledge requirement about the process to a minimum. The rest of this section is devoted to presenting a modified back-propagation algorithm based on above consideration, whereas numerical results are given in Section 8.5.

Denote the actual and desired process output vectors at time instant $t$ by $y_p^k(t)$ and $y_d(t)$ respectively, where $k$ indicates the $k$th iteration. The objective of the adaptive mechanism is to adjust adaptively the BNN parameters so as to minimize the cost function $E^{*k}$ defined in the time interval $[0,T]$ of interest by operating repeatedly the system

$$E^{*k} = \sum_{t=0}^{T} E^k(t) \tag{8.10}$$

where

$$E^k(t) = \frac{1}{2}\|y_d(t) - y_p^k(t)\| = \frac{1}{2}\sum_{l=1}^{m} (y_{d,l} - y_{p,l}^k)^2 \tag{8.11}$$

While the true gradient descent scheme should be performed in the weight space with respect to $E^{*k}$, corresponding to a batch training mode, many applications indicate the efficiency and feasibility of using an instantaneous gradient with respect to $E^k$, corresponding to a pattern training mode. This observation is of particular interest to us since it allows the parameters to be adjusted at each time instant. In the following, we will use $E^k$ as a basis for deriving the adaptive algorithm. For the sake of notational clarity, the superscript $k$ and time index $t$ will be omitted.

Suppose that the BNN network consists of three layers, an input layer, a hidden layer, and an output layer. The inputs of the BNN denoted by $v \in R^n$ receive the information from the outputs of the VSC network and thus, the input layer of the BNN has $n$ units. There are $n_h$ units at the hidden layer with logistic nonlinear activation functions. The outputs of the BNN denoted by $x \in R^m$ are the inputs of the process, therefore there are $m$ units at the output layer with linear activation functions.

Denote $w_{ji}^h (w_{ji}^o)$ as the weights connecting the $i$th unit at the input (hidden) layer to the $j$th unit at the hidden (output) layer. Following the gradient descent rule, the weight $w_{ji}^h$ and $w_{ji}^o$ can be updated with time by

$$\Delta w_{ji} = -\rho \cdot \frac{\partial E}{\partial w_{ji}} \tag{8.12}$$

By employing the chain rule, the $\frac{\partial E}{\partial w_{ji}}$ can be computed in the following way. For weights $w_{ji}^o$, we have

$$\frac{\partial E}{\partial w_{ji}} = \frac{\partial E}{\partial x_j} \cdot \frac{\partial x_j}{\partial w_{ji}} \tag{8.13}$$

where

$$\frac{\partial x_j}{\partial w_{ji}} = O_i^h \tag{8.14}$$

with $O_i^h$ being the nonlinear output of the ith unit at the hidden layer, and

$$\frac{\partial E}{\partial x_j} = \sum_{l=1}^{m} \left( \frac{\partial E}{\partial y_{p,l}} \cdot \frac{\partial y_{p,l}}{\partial x_j} \right) = -\sum_{l=1}^{m} (y_{d,l} - y_{p,l}) \cdot \frac{\partial y_{p,l}}{\partial x_j} \tag{8.15}$$

where $\frac{\partial y_{p,l}}{\partial u_j}$ is the partial derivative of the $l$th process output with respect to the $j$th process input. Thus, we have

$$\Delta w_{ji}^o = \rho \cdot \left[ \sum_{l=1}^{m} (y_{d,l} - y_{p,l}) \cdot \frac{\partial y_{p,l}}{\partial x_j} \right] \cdot O_i^h \tag{8.16}$$

Analogous to the notation of the standard back-propagation algorithm, equation (8.16) can be written as

$$\Delta w_{ji}^o = \rho \cdot \delta_j^o \cdot O_i^h \tag{8.17}$$

where

$$\delta_j^o = \sum_{l=1}^{m} (y_{d,l} - y_{p,l}) \cdot \frac{\partial y_{p,l}}{\partial x_j} \tag{8.18}$$

The update rule for hidden units has the same form as in the standard algorithm, i.e.

$$\Delta w_{ji}^h = \rho \cdot \delta_j^h \cdot v_j \tag{8.19}$$

and

$$\delta_j^h = f'(net_j) \cdot \sum_{k=1}^{m} \delta_k^o \cdot w_{kj}^o \tag{8.20}$$

where $f'$ is the derivative of the activation function with respect to the $net_j$, a weighted linear sum of the $j$ unit at the hidden layer.

Now we focus our attention on the adaptive algorithm for the output units given by equation (8.17). Denote $g_{lj} = \partial y_{p,l} / \partial u_j$ with $l, j = 1, 2, \cdots, m$. $G_{m \times m} = [g_{lj}]$ is referred to as the forward Jacobian of the process and is usually unknown in advance. Thus, a difficulty arises in applying (8.17) due to the requirement for the $G$ matrix. If the steady state gains relating the $l$th output with to the $j$th input at the operating point are known, they may be used to approximate the $G$ matrix near the operating point. However, by the assumption made in Section 8.2, no quantitative knowledge about the process is available. Thus, we have to rely upon some qualitative knowledge with some reasonable approximation.

To this end, the following observations are of particular interest. First, note that $g_{lj}$ can be absorbed into the learning rate in the following way:

$$\Delta w_{ji} = \left[ \sum_{l=1}^{m} \rho_{lj} \cdot (y_{d,l} - y_{p,l}) \right] \cdot O_i^h \tag{8.21}$$

where

$$\rho_{lj} = \rho \cdot \frac{\partial y_{p,l}}{\partial x_j} \tag{8.22}$$

Due to the nature of the gradient descent search, it is not only unnecessary but also impossible to assign the learning rate precisely apart from keeping it small enough to avoid divergence during the searching process. Second, the adaptive process in our system is assumed to be iterative and therefore it is hoped that any detrimental effects arising from the approximation could be compensated with increase in the iteration, a price which must be paid for any kind of learning system. Third, equation (8.21) indicates that the $\Delta w_{ji}$ are proportional to the weighted combination of $m$ tracking errors with the different credit assignments determined by $\rho_{lj}$. In other words, it shows how the present errors $y_{d,l} - y_{p,l}$ should be attributed to $w_{ji}$ responsible for producing $u_j$. In this sense, it seems more important to consider $\rho_{lj}$ as representing interaction effects within the control loops. However, if we notice that the error $y_{d,l} - y_{p,l}$, for which the $u_j$ should take principal responsibility, in fact, contains the information concerning the loop interaction, it may be reasonable to utilize only the error information explicitly related to $u_j$, i.e. $\rho_{lj} = 0$ if $l \neq j$.

In light of the above considerations, we propose the following extremely simple algorithm. By approximating the Jacobian matrix $G_{m \times m}$ with $G^*_{m \times m} = \text{diag}\{\text{sgn}[g_{11}], \text{sgn}[g_{11}], \cdots, \text{sgn}[g_{mm}]\}$ with sgn[·] denoting sign function, the algorithm at the output layer can be given by

$$\Delta w_{ji}^o = \rho \cdot sgn\,[g_{jj}] \cdot (y_{d,j} - y_{p,j}) \cdot O_i^h \tag{8.23}$$

whereas the algorithm at the hidden layer remains unchanged. We notice that sgn $[g_{jj}]$ are available by assumption. The feasibility of the above algorithm will be investigated in the next section by simulation studies.

## 8.5 Simulation results

The primary objective of the simulation is to investigate the self-organizing and self-learning behaviours of the proposed approach when applied to the multivariable blood pressure control problem as described in Appendix II. In particular, we are concerned with the issues of adaptive ability and convergence property of the learning system with respect to various situations, the achieved performances after learning, and the rule numbers produced during and after learning. Throughout this work, the following values were used. The time interval of interest contains 100 sampling points with sampling time being 30 s. Set-points for $CO$ and $MAP$ were set to be 20 ml/s and -10 mmhg changing from nominal values of 100 ml/s and 120 mmHg respectively. The maximum iteration number is 10. The other parameters used will be given accordingly.

1. *Achieved performance*

   By setting $\xi = 0.8$ and $t_s = 240$ s for the two controlled loops, learning rate in the BNN $\rho = 0.0001$,valid radius $\delta = 5$, the number of hidden units $n_h = 15$, and approximated Jacobian $G = \text{diag}[1,-1]$, the desired and actual output responses of the process after 10 iterations are shown in Figure 8.3, where (a), (b), and (c) correspond to the cases of EC, ES, and ECS type controllers. It is seen that all the results are quite acceptable although imperfect tracking is observed. However, we note that a relatively good transient performance was obtained by EC type, a good steady-state performance was achieved by ES type, and a good trade-off between transient and steady-state performances was given by ECS. These conclusions agree well with the conceptual considerations concerning the effects of E, C, and S actions on the performance. To give a quantitative evaluation of the performance, we define the following averaged squared sum of errors (ASSE) by

$$ASSE = \frac{1}{100}\sum_{i=1}^{100}\left[(y_{d,co}(i) - y_{p,co}(i))^2 + (y_{d,map}(i) - y_{p,map}(i))^2\right] \quad (8.24)$$

   Corresponding to Figure 8.3, the ASSE with EC, ES, and ECS types are listed in Table 8.1. It indicates that ECS type produced the best global performance. The numbers of units at the competitive layer in VSC net created by EC, ES, and ECS types are also listed in Table 8.1. We see that the EC type created the least number of rules (of course provided that the same valid radius is used for three cases.) This is due to the fact that fewer rules are needed to represent steady-state for the EC type in response to the step commands. Finally we compare the net complexity in terms of the number of adjustable weights. As illustrated in Table 8.1, the ECS type has the largest net size because of the larger number of inputs, whereas the EC type has the least net size due to fewer rule numbers. Thus, it can be concluded that while we chose the ECS type for its better global control performance, a price must be paid in terms of high complexity.

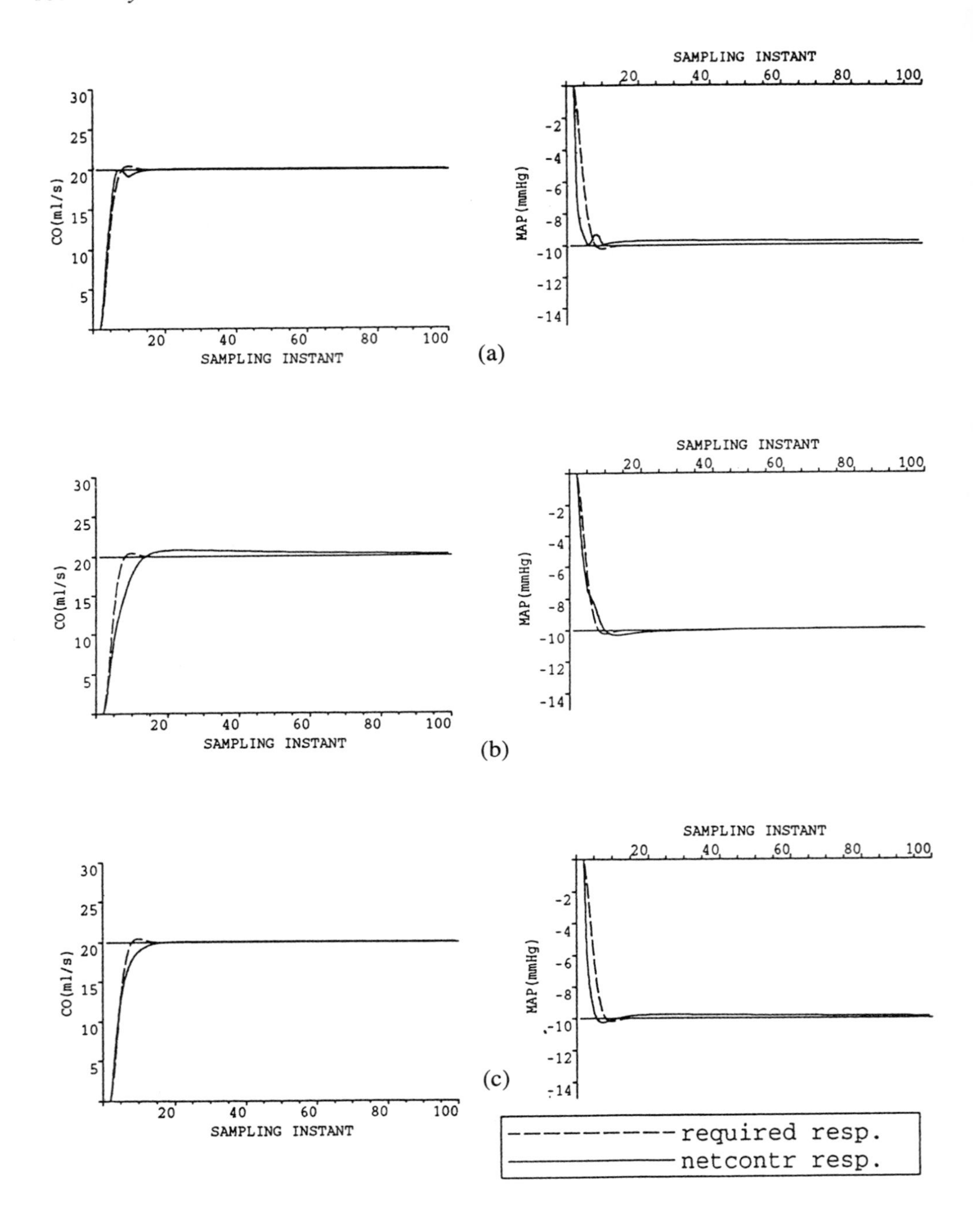


**Figure 8.3** Output responses of the process with EC, ES, and ECS types

**Table 8.1** Achieved performance and net size

| TYPE | ASSE | Rule no. | Weights in VSC | Weights in BNN | Total weights |
|---|---|---|---|---|---|
| EC | 1.0009 | 5 | 20 | 107 | 127 |
| ES | 1.5954 | 16 | 64 | 107 | 171 |
| ECS | 0.8675 | 10 | 60 | 137 | 197 |

2. *Adaptive ability*
This set of simulations is intended to study the learning or adaptive capacity of the proposed system with respect to the controlled environment, more specifically, to the variations of the process parameters and desired performance requirements, and to the noise measurements. To measure the performance, a normalized squared sum of errors (NSSE) is defined by

$$NSSE^k = \frac{SSE^k}{SSE_{\max}} \tag{8.25}$$

where

$$SSE^k = \sum_{i=1}^{100} \left[ (y_{d,co}(i) - y^k_{p,co}(i))^2 + (y_{d,map}(i) - y^k_{p,map}(i))^2 \right] \tag{8.26}$$

is the squared sum of error at the $k$th iteration and $SSE_{\max}$ is the largest $SSE^k$ chosen from all the $SSE^k$ produced by using different parameters within the same group at all the iterations from $k$=1 to $k$=10. As would be expected, $SSE_{\max}$ usually occurs at the first iteration.

Because we are concerned with the effects of the controlled environment, it makes sense only when the controller parameters remain fixed with respect to all the situations. In this case, $\rho = 0.0001$, $\delta = 5$, $n_h = 15$, and $G = \text{diag}[1,-1]$ were used throughout this set of simulations. Figures 8.4 and 8.5 show the results of NSSE versus the iteration number in the case of the ECS type when the process gains (Figure 8.4) and time constants (Figure 8.5) were changed from their nominal values to the values indicated in the figures. It appears that the learning system with a set of fixed controller parameters is able to handle the process with process parameter variations of up to 20%, in the sense that all the NSSE after a few iterations became small and stable and remained close to each other. This implies that the system possesses very good robustness stemming mainly from

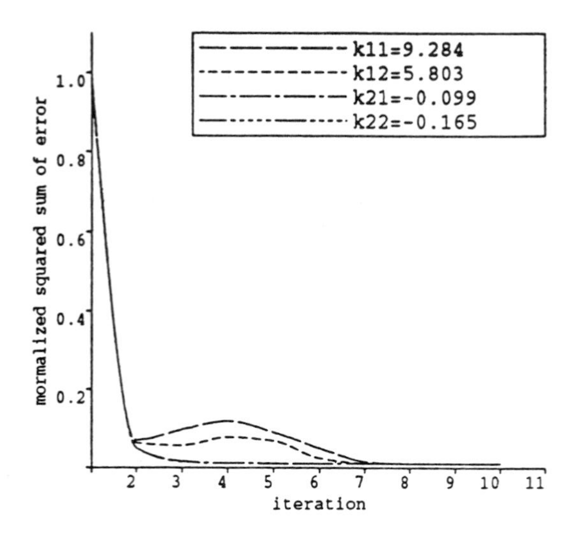


**Figure 8.4** Adaptive ability to different process gains

the learning capability of the system. The same conclusion can be made with respect to the different performance requirements, as depicted in Figure 8.6, where the desired responses are specified as follows:

(a) $\xi_{co} = 0.7$ $t_{s,co} = 240$ s $\xi_{map} = 0.8$ $t_{s,map} = 240$ s
(b) $\xi_{co} = 0.8$ $t_{s,co} = 240$ s $\xi_{map} = 0.7$ $t_{s,map} = 240$ s
(c) $\xi_{co} = 0.8$ $t_{s,co} = 240$ s $\xi_{map} = 0.8$ $t_{s,map} = 240$ s
(d) $\xi_{co} = 0.9$ $t_{s,co} = 400$ s $\xi_{map} = 0.7$ $t_{s,map} = 240$ s

To examine the learning ability in a noise-contaminated environment, a set of simulations was carried out in which the outputs of the process were corrupted by random noise with a uniform distribution and therefore both controller inputs and adaptive inputs are contaminated by noise. Figure 8.7 shows the outputs and inputs of the EC type controller with a noise amplitude of 5% at the set-points. It can be seen that the responses are satisfactory. We notice that because of the averaging effects of the VSC on its noise inputs, the controller outputs are relatively less sensitive to the noise compared with the process outputs. Figure 8.8 illustrates the convergence behaviour of the system under the different noise amplitudes. In all cases, the SSE decreases with iteration and reaches a stable value.

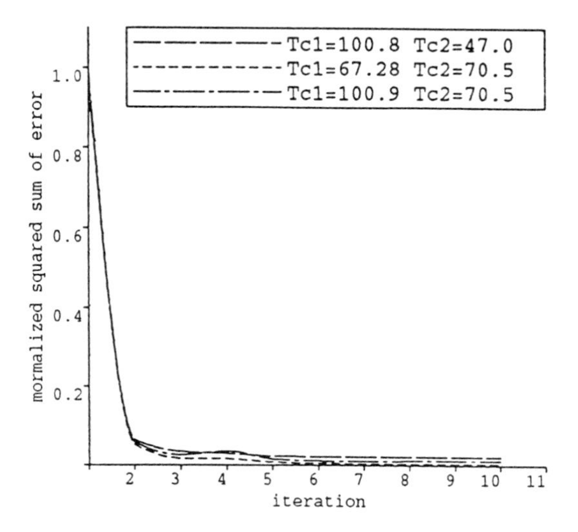


**Figure 8.5** Adaptive ability to different process time-constants

3. *Convergence property*
   This group of simulations focused on examining the convergence properties of the adaptive systems with various controller or net parameters while keeping the process parameters fixed. More specifically, we are concerned with the question of how these parameters influence the convergence process in terms of the resultant NSSE. The parameters of interest include the number of hidden units and learning rate in BNN, the valid radius in VSC, and Jacobian approximation in the adaptive algorithm. In what follows, the ECS type will be adopted in all the simulation studies.
   *The number of hidden units.* It is generally known that too few or too many hidden units in the BNN will typically result in either a poor convergence in the training stage or a poor generalization in the operational stage. Unfortunately, there is no generic method available to specify this parameter for a specific problem. The main concerns in our case are the convergence rate during the learning process and the achieved performance after the learning. Figure 8.9 shows the variations of the NSSE with iteration number under three cases. It can be seen that similar convergence behaviour is possessed by all the cases. However, a bigger initial SSE was observed in the case of 20 and 5 units and the smallest SSE was obtained with 5 units, suggesting that it is reasonable to use a relatively small number of hidden units for the current problem.

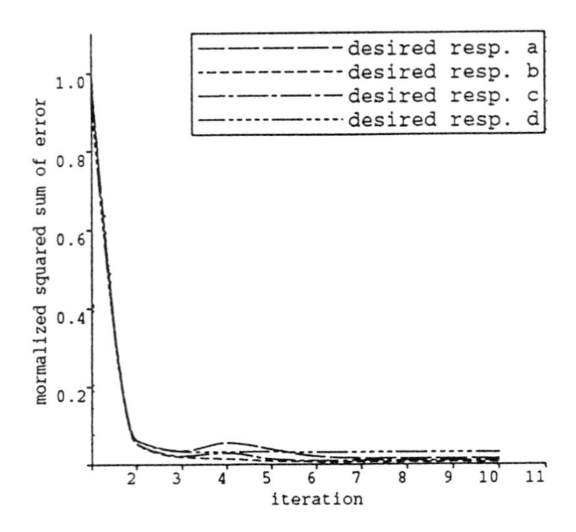


**Figure 8.6** Adaptive ability to various desired responses

*Learning rate.* Figure 8.10 illustrates the simulation results with different learning rates $\rho$. As would be expected, the system with a smaller $\rho$ produced a bigger initial SSE and a slower convergence speed. This is indicated clearly in the case of a learning rate of 0.0001. While a higheer learning rate is desired for the fast convergence, care must be taken not to chose it so large as to cause divergence.

*Jacobian approximation.* At first glance, one may be in doubt about the feasibility of the approximation procedure made in developing the adaptive algorithm where the required Jacobian is approximated by a diagonal matrix with elements being 1 or -1. However, remarkably good performance has been achieved with such an extremely simple algorithm. In fact, all the simulations reported above employed this approximation. In order to verify how the performance would be improved if the exact Jacobian or steady-state gain matrix is used, we carried out a comparative study. With the nominal process parameter values, we derived the following relative gain matrix:

$$G = \begin{bmatrix} 1 & 0.84 \\ -0.16 & -1 \end{bmatrix} \tag{8.27}$$

where 0.84 and -0.16 are relative gains representing the interactive effects between the variables. Figure 8.11 gives the simulation results compared with the case of diagonal approximation. While a similar convergence tendency was obtained, a better performance (a small NSSE) was achieved in the case of the

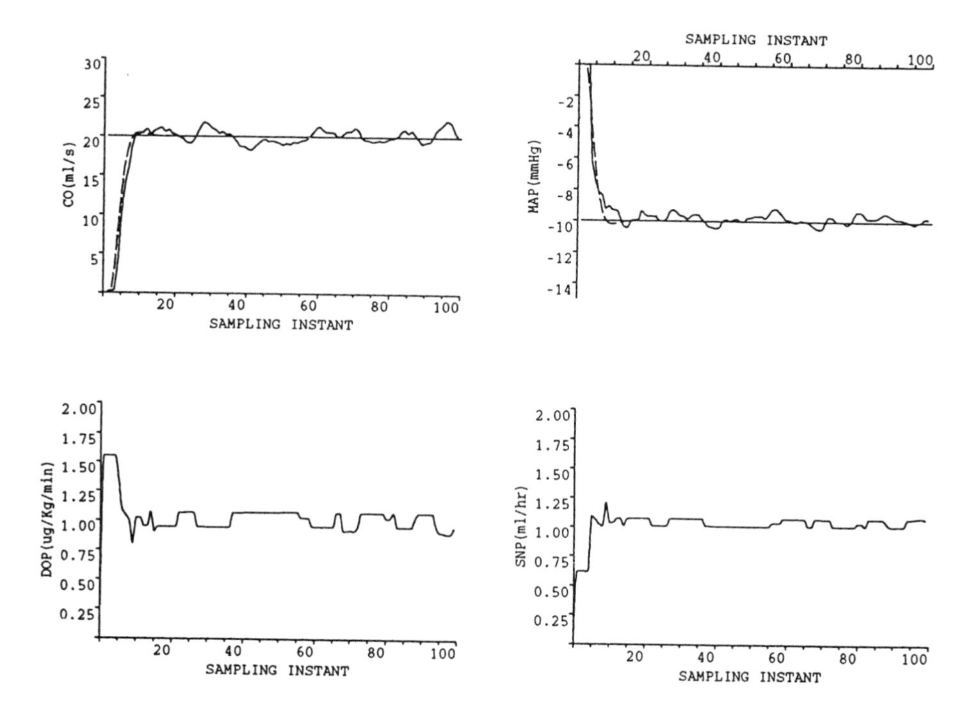


**Figure 8.7** Outputs and inputs of the process with EC type under noise measurements

known gain matrix. Thus, we can conclude that although the approximation scheme may be feasible, it would be beneficial to make use of as much prior knowledge about the process as available.

*Valid radius.* The valid radius $\delta$ is an important parameter controlling the self-organizing process in the VSC net. The bigger $\delta$ is, the more units the VSC creates, the more rules in the rule-base, and therefore, the more input patterns to the input of the BNN. Because the valid radius is designed mainly to accomplish the partitioning of the input space, we expect that the control performance would be not very sensitive to this parameter. Figure 8.12 gives the simulation results using three different radii with the values indicated in the figure. It can be seen that the three curves are almost identical, thereby verifying our expectation. However, the number of rules created were substantially different as shown in Figure 8.13. It is interesting to note that with increase of the iteration, the number of rules tends to a stable value, as did the NSSE and therefore it can also be considered as a convergence indicator.

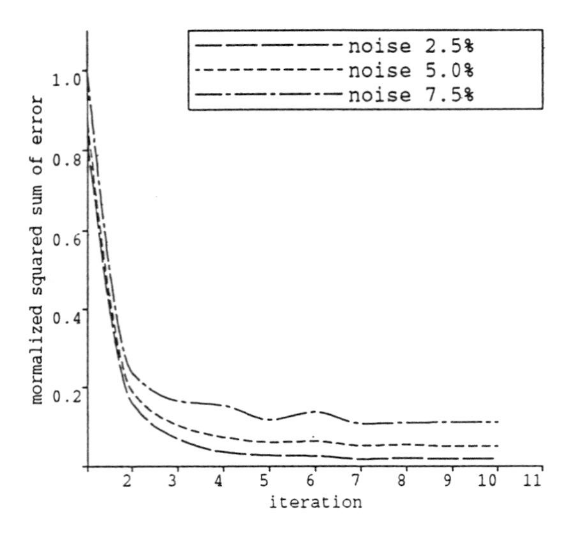


**Figure 8.8** Convergence property under different noise amplitudes

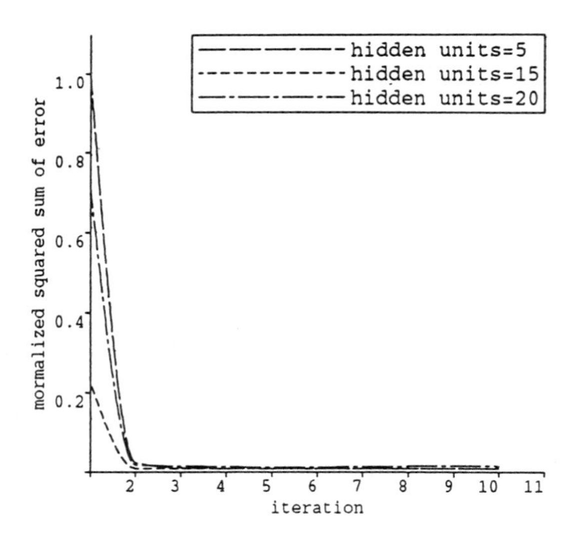


**Figure 8.9** Convergence property with different numbers of hidden units

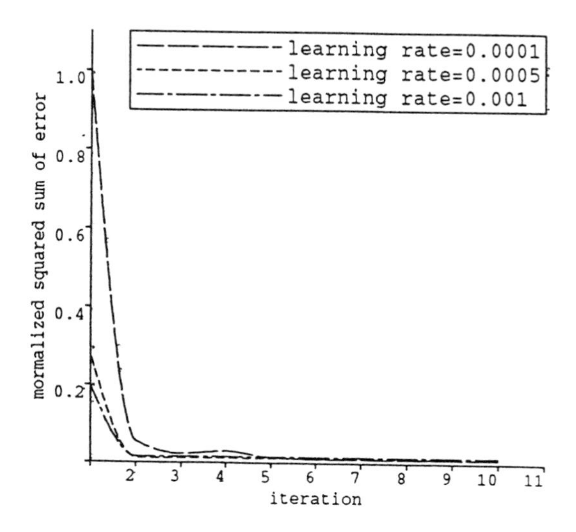


**Figure 8.10** Convergence property with different learning rates

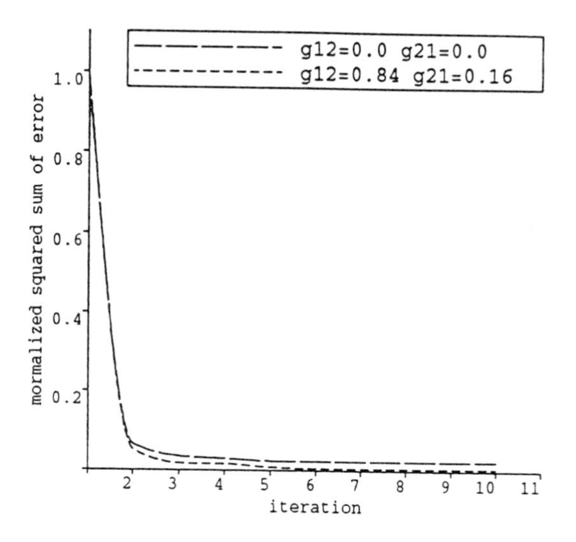


**Figure 8.11** Convergence property with different Jacobian approximations

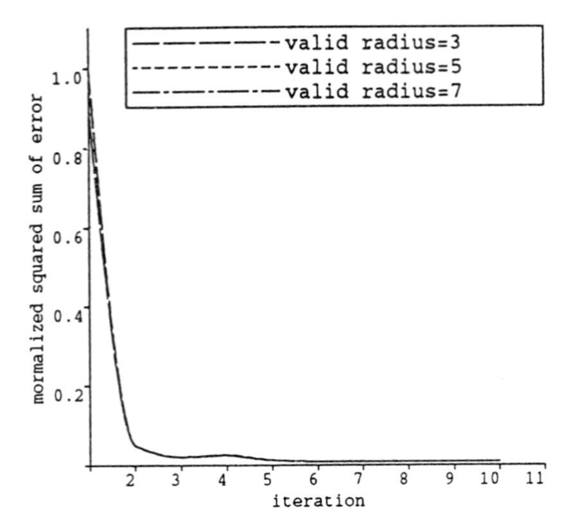


**Figure 8.12** Convergence property with different valid radii

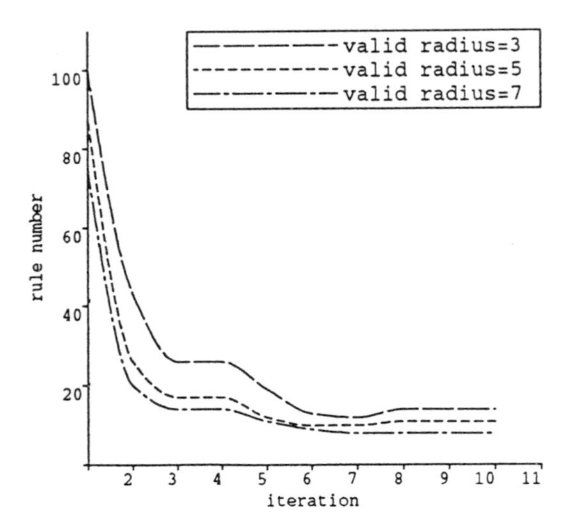


**Figure 8.13** Rule numbers created by different valid radii

## 8.6 Summary

We have presented a novel approach to constructing systematically a self-organizing and self-learning multivariable fuzzy controller. The proposed controller is built with hybrid neural networks consisting of a variable-structure competitive network and a standard BNN network. We have developed the corresponding self-organizing and adaptive learning algorithms. The system proposed in this chapter is an outcome of incorporating the paradigm of knowledge-based fuzzy controllers with neural networks. In particular, we view the neural network-based paradigm not only from a computational standpoint, but also from knowledge representation and processing aspects, thereby taking advantage of the computational efficiency and adaptive capability of the network, enabling one to maintain the clarity of the rule-based paradigm and thus providing an explicit explanation facility for the BNN network. This characteristic may be of practical interest when one is concerned with what functions a black-box-like net-based controller has performed. This can be done easily by inspecting the resultant rule-base because the behaviour of the BNN has been summarized by a set of corresponding rules. The other features of the approach include the following: (a) the scheme is very simple, there being no separate training stage involved, nor an identification procedure necessary; (b) very little prior knowledge about the process is required in the implementation of the algorithm but at the price of iterative learning; (c) the resultant rule-base may be used as a basis for designing a much simpler fuzzy controller for dealing with the same controlled system, possibly in a look-up table manner; (d) the PID-like controller input provides the possibility of applying the proposed algorithm directly to widely existing PID control systems without much extension.

Extensive simulation studies on the problem of multivariable blood pressure control have not only demonstrated the feasibility of the proposed system, but also offered some insights into a better understanding of many aspects possessed by the system structure. In particular, we have investigated the adaptive ability and learning convergence properties with respect to a wide range of situations subject to different net parameters and controlled environments. It has transpired that this relatively simple scheme has a remarkable capability of dealing with such a relatively complex multivariable system having strong interaction between variables.

CHAPTER 9

# CPN network-based fuzzy controller: explicit representation and self-construction of rule-bases

*By structurally mapping a simplified fuzzy control algorithm (SFCA) into a modified counterpropagation network (CPN) network, a simple but efficient CPN-based fuzzy controller is obtained with the characteristics of explicit representation, self-construction of rule-bases, and fast learning speed.*

## 9.1 Introduction

While the previous three chapters have focused primarily on the functional mapping between an approximate reasoning algorithm and a neural network approach, this chapter is concerned mainly with the structural mapping between those two paradigms. However, the basic issues we address remain the same regardless of the methods adopted, that is, rule-base acquisition, computational representation, and approximate reasoning. The objective is to deal with these issues under a unified framework of neural networks in such a way that the rule-base can be constructed automatically, also the fuzzy reasoning mechanism can be implemented easily, and the required prior knowledge about the controlled environment should be as little as possible. We present here a novel approach capable of fulfilling the objectives specified above. In particular, we describe a simple and efficient scheme capable of self-organizing and self-learning the required control knowledge in a real-time manner under multivariable controlled environments. The starting point of the approach is to structurally map a simplified fuzzy control algorithm (SFCA) (Nie 1987, 1989) into a counterpropagation network (CPN) developed by Hecht-Nielsen (1987, 1988, 1990). The CPN is of a very simple structure, but often overlooked, and is used here in such a way that the control knowledge is explicitly represented in the form of connection weights of the nets. Then, by introducing a valid radius into the Kohonen layer and providing an on-line learning teacher to the Grossberg layer, the control rule-base, initially empty, is gradually self-organized and self-constructed to achieve fulfilment of the prespecified performance requirements. Finally, the approximate reasoning is carried out by replacing a winner-take-all competitive scheme with a soft matching cooperative strategy.

The next section introduces a simplified fuzzy control algorithm suitable for sensor database control environments. Section 9.3 is devoted to mapping the fuzzy control algorithm to the CPN network relevant to the issues of representation and reasoning. The underlying principles concerning self-organizing and self-learning are described in Section 9.4. As a demonstration example, a problem of multivariable control of blood pressure is studied in Section 9.5 and the chapter ends with some concluding remarks in Section 9.6.

## 9.2 Simplified fuzzy control algorithms (SFCA)

The operation of a fuzzy controller typically involves the following main stages within one sampling instant: sampling the measured inputs, fuzzifying the nonfuzzy inputs into fuzzy sets, inferencing the current controller fuzzy outputs, and finally defuzzifying the inferred fuzzy outputs into nonfuzzy values and sending them to the process being controlled. Among these, the reasoning stage usually comprises three further substages, namely, computing the matching degrees between the current fuzzy inputs with respect to each rule's IF part, determining which rules should be fired, and combining these fired rules' THEN parts with different strengths into final fuzzy sets. It can be observed that it is the incompatible property of the numerical (or nonfuzzy) control environment with the linguistic (or fuzzy) reasoning algorithm that makes the operation complex in terms of requiring two interfaces to connect the external nonfuzzy environment with a fuzzy inference engine which is developed normally within a fuzzy environment. In our view, it is possible to simplify the above process by taking the nonfuzzy property regarding the input–output of the fuzzy controller into account. We have derived a very simple but efficient MISO fuzzy control algorithm SFCA which consists of only two main steps, pattern matching and weighted averaging, thereby eliminating the necessity for fuzzifying and defuzzifying procedures (Nie 1987, 1989). The basic idea is to make the approximate reasoning strategy usually derived for linguistic variables suitable directly for use with numerical variables by introducing notions of pattern and pattern matching. The algorithm can be extended to the MIMO case in a straightforward way and is described below.

*Input and output.* Assume that the controlled process is multivariable with $m$ inputs and $m$ outputs. The inputs to the fuzzy controller are various combinations of control error, change-in-error, and sum of error. Suppose that the measured control error at time $lT_s$ with $T_s$ sampling period is denoted as $e_{c1}(lT_s)$, $e_{c2}(lT_s)$, $\cdots$, $e_{cm}(lT_s)$ corresponding to $m$ process outputs, i.e. $e_{ci}(lT_s)$ =

$r_i(lT_s) - y_{p,i}(lT_s)$, where $r$, $y_p \in R^m$ are set-point and process output vectors respectively. Each $e_c(lT_s)$ is extended into three elements: $e_c(lT_s)$, $c_c(lT_s)$, and $s_c(lT_s)$, where

$$c_{cl}(lT_s) = e_c(lT_s) - e_c((l-1)T_s) \tag{9.1}$$

is the change-in-error and

$$s_c(lT_s) = \sum_{i=1}^{l} e_c(iT_s) \tag{9.2}$$

is the sum of error. Thus, in light of different combinations, three possible controller input modes can be constructed, i.e.

1. Input variables: $e_{c1}, c_{c1}, e_{c2}, c_{c2}, \cdots, e_{cm}, c_{cm}$
2. Input variables: $e_{c1}, s_{c1}, e_{c2}, s_{c2}, \cdots, e_{cm}, s_{cm}$
3. Input variables: $e_{c1}, c_{c1}, s_{c1}, e_{c2}, c_{c2}, s_{c2}, \cdots, e_{cm}, c_{cm}, s_{cm}$

We refer to the input types 1, 2 and 3 as EC, ES, and ECS respectively. In what follows, it is assumed that the controller mode is one of these types and is composed of $n$ variables, each of which is denoted by $u_i$. The output of the fuzzy controller consists of $m$ variables, each of which is denoted by $v_k$. Thus, a $n \times m$ controller is used to control a $m \times m$ process with $n > m$ in general.

*Rule pattern and input pattern.* Assume that there are $P$ rules in the rule-base, each of which has the form:

$$IF\ U_1\ is\ A_1^j\ AND\ U_2\ is\ A_2^j\ AND\ \cdots\ AND\ U_n\ is\ A_n^j$$
$$THEN\ V_1\ is\ B_1^j\ AND\ V_2\ is\ B_2^j\ AND\ \cdots\ AND\ V_m\ is\ B_m^j$$

where $U_i$ and $V_k$ are linguistic variables corresponding to the numerical variables $u_i$ and $v_k$, $A_i^j$ and $B_k^j$ are fuzzy subsets representing some linguistic terms and defined on the corresponding universes of discourse $\overline{U}_i$ and $\overline{V}_k$ which are assumed to be compact on $R$. Fuzzy subsets $A_i^j$ and $B_k^j$ are characterized by the corresponding membership functions $A_i^j(u_i) : \overline{U}_i \rightarrow [0,1]$, and $B_k^j(v_k)$: $V_k \rightarrow [0,1]$. More precisely, let $A_i^j$ and $B_k^j$ be normalized fuzzy subsets whose membership functions are defined uniquely as

$$A_i^j(u_i) = \begin{cases} 1 - \left[ \dfrac{|M_{u,i}^j - u_i|}{\delta_{u,i}^j} \right] & \text{if } |M_{u,i}^j - u_i| \le \delta_{u,i}^j \\ 0 & \text{if } |M_{u,i}^j - u_i| > \delta_{u,i}^j \end{cases} \tag{9.3}$$

$$B_k^j(v_k) = \begin{cases} 1 - \left[ \dfrac{|M_{v,k}^j - v_k|}{\delta_{v,k}^j} \right] & \text{if } |M_{v,k}^j - v_k| \le \delta_{v,k}^j \\ 0 & \text{if } |M_{v,k}^j - v_i| > \delta_{v,k}^j \end{cases} \qquad (9.4)$$

where $M_{u,i}^j \in \bar{U}_i$, $M_{v,k}^j \in \bar{V}_k$, $\delta_{u,i}^j > 0$, and $\delta_{v,k}^j > 0$.

Observe that each of the membership functions in equations (9.3) and (9.4) is characterized by only two parameters, $M_{u,i}^j$ and $\delta_{u,i}^j$, or $M_{v,k}^j$ and $\delta_{v,k}^j$ with the understanding that $M_{u,i}^j$ ($M_{v,k}^j$) is the centre element of the support set of $A_i^j$ ($B_k^j$), and $\delta_{u,i}^j$ ($\delta_{v,k}^j$) is the half width of the support set. Hence $A_i^j$ and $B_k^j$ may be expressed as

$$A_{u,i}^j = (M_{u,i}^j, \delta_{u,i}^j); \quad B_{v,k}^j = (M_{v,k}^j, \delta_{v,k}^j) \qquad (9.5)$$

By using the above notation, the $j$th rule can be written as

*IF* $(M_{u,1}^j, \delta_{u,1}^j)$ *AND* $(M_{u,2}^j, \delta_{u,2}^j)$ *AND* . . . *AND* $(M_{u,n}^j, \delta_{u,n}^j)$

*THEN* $(M_{v,1}^j, \delta_{v,1}^j)$ *AND* $(M_{v,2}^j, \delta_{v,2}^j)$ *AND* . . . *AND* $(M_{v,m}^j, \delta_{v,m}^j)$

In the rest of this section and the next subsection we will focus on the IF part of the above rule, and leave the THEN part to be treated latter on. Let input space $\Omega = (\bar{U}_1 \times \bar{U}_2 \times \ldots \times \bar{U}_n) \in R^n$ be a compact product space, and $M_u^j = (M_{u,1}^j, M_{u,2}^j, \ldots, M_{u,n}^j) \in \Omega$ and $\Delta_u^j = (\delta_{u,1}^j, \delta_{u,2}^j, \ldots, \delta_{u,}\, n^j)$ be two $n$-dimensional vectors. Then the *condition* part of the $j$th rule may be viewed as creating a subspace $\Omega^j \in \Omega$ or a hypercube whose centre and radius are $M_u^j$ and $\Delta_u^j$ respectively. Thus, the *condition* part of the $j$th rule can be simplified further to "IF $M\Delta_u$(j)", where $M\Delta_u(j) = (M_u^j, \Delta_u^j)$. Similarly, $n$ current inputs $u_{0i} \in \bar{U}_i$ ($i$ = 1, 2, . . ., $n$), with $u_{0i}$ being a singleton, can also be represented as a $n$-dimensional vector $u_0$ in $\Omega$.

It is helpful to view the IF part of a rule and a measured input as patterns to be called a *rule pattern* and an *input pattern* respectively. Geometrically, a rule pattern can be visualized as consisting of a set of points (vectors) centred at $M_u^j$ with a neighbourhood defined by $\Delta_u^j$ with constraints imposed by the corresponding membership grades. $P$ rule patterns partition $\Omega$ into $P$ subspaces which are typically overlapped to some degree along the boundaries due to the effect of fuzziness. In contrast, a measured input is just a determined point situated in the same space as are the rule patterns.

*Pattern matching.* The fuzzy control algorithm can be considered to be a process in which an appropriate control action is deduced from a current input and $P$ rules according to some prespecified reasoning algorithms. We split the whole reasoning procedure into two phases: pattern matching and weighted averaging. The first operation deals with the IF part for all rules, whereas the second involves an operation on the THEN part for the fired rules. From the pattern concept introduced above, we need to compute the matching degrees between the current input pattern, a point in $\Omega$, and each rule pattern, a set of points in $\Omega$. Because the two patterns are now formulated numerically in the same manner and are interpreted geometrically in the same space, it is straightforward to adopt some metrical concepts to measure the similarity of the two patterns. There are at least two approaches to performing this task, based on the notions of *volume ratio* or *relative distance*. For the sake of later use in this chapter, we present only the distance algorithm as follows.

Denote the current input by $u_0 = (u_{01}, u_{02}, \ldots, u_{0n})$. Then the matching degree denoted by $S^j \in [0, 1]$ between $u_0$ and the $j$th rule pattern $M\Delta_u(j)$ can be measured by the complement of the corresponding relative distance given by

$$S_j = 1 - D^j(u_0, M\Delta_u(j)) \tag{9.6}$$

where $D^j(u_0, M\Delta_u(j)) \in [0,1]$ denotes relative distance from $u_o$ to $M\Delta_u(j)$. $D^j$ can be specified in many ways and three computational definitions are given below.

*Relative Euclidean Distance:*

$$D^j_E(u_0, M\Delta_u(j)) = \begin{cases} \dfrac{\|M^j_u - u_0\|}{\|\Delta^j_u\|} & \text{if } \|M^j_u - u_0\| \le \|\Delta^j_u\| \\ 1 & \textit{otherwise} \end{cases} \tag{9.7a}$$

where $\|\cdot\|$ denotes the Euclidean norm.

*Relative Hamming Distance:*

$$D^j_H(x_0, M\Delta_u(j)) = \begin{cases} \dfrac{\sum_{i=1}^{n} |M^j_{u,i} - u_{0i}|}{\sum_{i=1}^{n} \delta^j_{u,i}} & \text{if } \sum_{i=1}^{n} |M^j_{u,i} - u_{0i}| \le \sum_{i=1}^{n} \delta^j_{u,i} \\ 1 & \textit{otherwise} \end{cases} \tag{9.7b}$$

*Relative Maximum Distance:*

$$D^j_M(x_0, M\Delta_u(j)) = \begin{cases} \underset{1 \le i \le n}{Max} \left( \dfrac{|M^j_{u,i} - u_{0i}|}{\delta^j_{u,i}} \right) & \text{if for } \textit{all } i \quad |M^j_{u,i} - u_{0i}| \le \delta^j_{u,i} \\ 1 & \textit{otherwise} \end{cases} \tag{9.7c}$$

It is evident that, from equations (9.6) and (9.7), if $u_0$ and $M\Delta_u(j)$ are fully matched, i.e. $u_0$ is exactly the same as the centre vector $M^j_u$, then $D^j = 0$, leading to the matching degree $S^j$ being 1. In contrast, if they are completely unmatched, i.e. $u_0$ is on or outside the boundary of $M\Delta_u(j)$ determined by the corresponding metric, then $D^j = 1$ and thus, $S^j = 0$. Otherwise $0 < D^j < 1$ and $0 < S^j < 1$, indicating a partial matching. It is worth pointing out that the subspaces created by the above distance matrices are of different shapes. More specifically, while $D_E$ and $D_H$ produce spherical and diamond supports respectively, the $D_M$ gives a cube support. Although these different algorithms do not have significant effects on the reasoning performance, the $D_H$ and $D_M$ are simpler in computation and furthermore the $D_M$ is particularly suitable for hardware implementation.

*Weighted averaging.* Recall that under the unique definition of the membership function, each rule can be expressed as

$$IF\ M\Delta_u(j)\ THEN\quad (M^j_{v,1}, \delta^j_{v,1})\ AND\ \ldots\ AND\ (M^j_{v,m}, \delta^j_{v,m})$$

In what follows, we assume that $\delta^j_{v,k}$ are identical with respect to all $k$ and $j$. Suppose that a current input pattern $u_0$ and a specific rule pattern $j$ are given. If the matching degree $S^j = 1$, the deduced control values should be $v_k = M^j_{v,k}$, $k = 1, 2, \ldots, m$. This conclusion implies the utilization of the *Maximum Membership Decision* scheme. However, it is also identical to the scheme referred as to *Centre of Gravity* (COG) if we notice that the membership functions are assumed to be symmetrical about their centres. If $S^j = 0$, on the other hand, the $j$th rule has no contribution to the final controller output. Otherwise, $0 < S^j < 1$ and there are more than one rule contributing to the control values.

Suppose that for a specific input $u_0$ and $P$ rules, after the matching process is completed, there exist $Q$ matching degrees satisfying $0 < S < 1$ and they are relabelled as $S^1, S^2, \ldots, S^Q$ with corresponding $Q$ groups of centres of the THEN parts denoted by

$$\left\{M^1_{v,1}, M^1_{v,2}, \cdots, M^1_{v,m}\right\} \left\{M^2_{v,1}, M^2_{v,2}, \cdots, M^2_{v,m}\right\} \ \ldots \ \left\{M^Q_{v,1}, M^Q_{v,2}, \cdots, M^Q_{v,m}\right\}$$

Then the $k$th component of the deduced control action $v_k$ is given by

$$v_k = \frac{\sum_{q=1}^{Q} S^q \cdot M^q_{v,k}}{\sum_{q=1}^{Q} S^q} = \sum_{q=1}^{Q} \bar{S}^q \cdot M^q_{v,k} \tag{9.8}$$

where

$$\bar{S}^q = \frac{S^q}{\sum_{q=1}^{Q} S^q}$$

It can be seen that equation (9.8) gives a weighted averaging value with respect to the fired rules' THEN parts. How large a percentage a specific rule contributes to the global value is determined by the corresponding matching degree. Because only the centres of the THEN parts of the fired rules are utilized and they are the only elements having the maximum membership grade 1 on the corresponding support sets, the algorithm can be understood as a modified maximum membership decision scheme in which the global centre is calculated by the COG algorithm. However, notice that compared with the traditional defuzzification COG algorithm where the centre is determined by all the elements in $\bar{V}_k$, here it is computed only by the local centres of $Q$ fired rules. Because symmetrical membership functions with identical widths are assumed, the effect produced by omitting the width attached to the THEN part can be surely neglected, thereby suggesting that algorithm (9.8) is justified, also the widths in the THEN parts can be removed from the original rules, and accordingly the rule form can be further simplified to

$$IF\ \ M\Delta_u(j)\quad THEN\quad M_v^j \tag{9.9}$$

where $M_v^j = [M_{v,1}^j, M_{v,2}^j, \ldots, M_{v,m}^j]$ is a centre value vector of the THEN part.

## 9.3 Representation and reasoning by CPN

### 9.3.1 CPN network

By combining a portion of the self-organizing map of Kohonen and the outstar structure of Grossberg, Hecht-Nielsen developed a new type of neural network named counterpropagation network (CPN) (Hecht-Nielsen 1987, 1988, 1990) Functionally, CPN is designed to approximate a continuous function $f: A \in R^n \rightarrow R^m$, defined on a compact set $A$ by means of a set of samples $(\mu^s, \nu^s)$ with $\mu^s$ vectors being randomly drawn from $R^n$ in accordance with a fixed probability density function $\rho$. The trained CPN functions as a statistically optimal self-adapting look-up table. Possible applications of CPN include pattern recognition, function approximation, statistical analysis, and data compression (see Hecht-Nielsen 1988).

Figure 9.1 shows a schematic of the *forward-only* version of the CPN which will be used in this chapter. It consists of an input layer, a hidden Kohonen layer, and a Grossberg output layer with $n$, $N$, and $m$ units respectively. In what follows, the forward algorithm used during normal operation of the CPN is presented, whereas the backward algorithm used during training will be described in the next section. Denoting the input vector and output vectors at the Kohonen and Grossberg layers by $\mu = [\mu_1, \mu_2, \ldots, \mu_n]^T$, $\zeta = [\zeta_1, \zeta_2, \ldots, \zeta_N]^T$, and $\nu = [\nu_1, \nu_2, \ldots, \nu_m]^T$ respectively, the *single winner* forward algorithm of the CPN in regard to a particular input $\mu_0$ is outlined as follows:

1. Determine the winner unit $J$ at the Kohonen layer competitively according to the distances of weight vector $\omega^j(t)$ with respect to $\mu_0$

$$D(\omega^J, \mu_0) = \min_{j=1,N} D(\omega^j, \mu_0) \tag{9.10}$$

where $\omega^j = [\omega_1^j, \omega_1^j, \cdots, \omega_n^j]$ is the weight vector connecting $n$ input units to

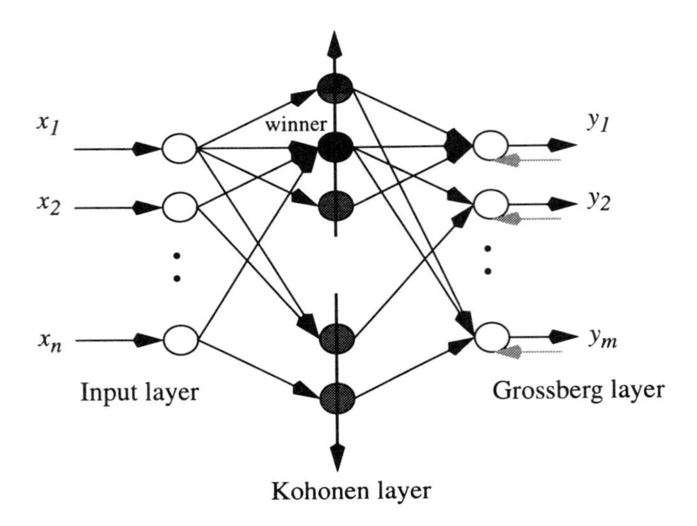


**Figure 9.1** The forward-only counterpropagation network

the $j$th unit at the Kohonen layer.

2. Calculate the outputs $\zeta^j \in \{0,1\}$ of the Kohonen layer by a winner-take-all rule

$$\zeta^j = \begin{cases} 1 & \text{if } j = J \\ 0 & \textit{otherwise} \end{cases} \tag{9.11}$$

3. Compute the outputs of the Grossberg layer by

$$\nu_k = \sum_{j=1}^{N} \zeta^j \cdot \pi_k^j \tag{9.12}$$

where $\pi^j = [\pi_1^j, \pi_1^j, \ldots, \pi_m^j]$ is the weight vector connecting the $j$th Kohonen unit to $m$ Grossberg output units and $k = 1, 2, \ldots, m$.

To improve the mapping approximation accuracy, the trained CPN can also be operated in a multi-winner mode or an *interpolation* mode. The basic idea is that more than one Kohonen unit can win the competition with respect to the current input $\mu_0$, and therefore, the computed net output is a combination of look-up table entries associated with the winning Kohonen units. A problem arises as to how to split the unit output signal ($\zeta^j = 1$) which was previously assigned to a single winner into several portions distributed to each winner. Hecht-Nielsen (1988) suggested that the win criterion can be based either on distance measurement between the current input and winning weights or on the basis of a fixed number of Kohonen units winning. As pointed out by Hecht-Nielsen (1990), this partitioning scheme used to split the $\zeta$ output signal is one of the major open research issues in connection with the CPN network.

### 9.3.2 Equivalence between SFCA and CPN

As mentioned in Section 9.1, we are seeking structural mapping from the SFCA algorithm onto the CPN network such that control knowledge represented formerly in rule form can be represented by the topology of the CPN and associated weights, and the reasoning algorithm could be carried out by operating the network directly. By carefully examining these two paradigms, we have found that there exist some striking similarities between the SFCA and CPN which provide a basis for performing our objective.

The first aspect of similarity involves the issue of knowledge representation. More specifically, after training is terminated the CPN network has been trained and arranged in such a way that $N$ associations $(\omega^j, \pi^j)$ can function as a look-up table. In other words, the knowledge acquired via training is entirely represented by associated weights $\omega^j$ and $\pi^j$. Furthermore, as will be explored in the next section, the learned vectors $\omega^j$ are approximately equi-probable with respect to $\mu$ vectors drawn from $A$ and each $\omega^j$ is in fact a representative of a set of vectors surrounding it. It becomes very clear that, by being analogous to the simplified rule form (9.9), each CPN association $(\omega^j, \pi^j)$ can be expressed as a rule:

$$IF \quad \omega^j \quad THEN \quad \pi^j \tag{9.13}$$

Equivalently, each Kohonen unit defines a rule by regarding the weight vectors connecting to and emanating from that unit as IF and THEN parts respectively. However, the difference between rules (9.9) and (9.13) lies in the specification of the width related to the IF part of the rule. While the width is explicitly present in (9.9), it is in fact implicitly embedded in (9.13) and becomes effective during the operation stage in a nearest-neighbour sense. Recall that $M_u^j$ in rule (9.9) is a centre vector with maximum membership grade 1 and plays the same role as $\omega^j$ in (9.13). In this view, each rule in the form of (9.9) can be directly mapped into a Kohonen unit with $M_u^j$ and $M_v^j$ being corresponding weight vectors.

The similarity between these two systems can be made even clearer by exploring the algorithmic aspect in CPN and reasoning process in SFCA. Both of them are intended to provide an approximate output $\nu$ or $v$ by operating the existing knowledge $(\omega^j, \pi^j)$ or $(M_u^j, M_v^j)$ with respect to the current input $\mu$ or $u$. Comparing equations (9.6), (9.7), and (9.8) in SFCA with equations (9.10), (9.11), and (9.12) in CPN, it can be concluded that the matching degree stage in SFCA is functionally similar to the competition process in CPN, and also the weighted average in SFCA corresponds to the computation of the output at the Grossberg layer. An interesting and important analogy is the correspondence between matching degree $S^j$ in SFCA and the activation value $\zeta^j$ at the Kohonen layer in CPN. $S^j$ usually takes a value in the range of [0,1], indicating a graded or soft matching and leading to an approximate output by averaging all the $M_v^j$ with $S^j > 0$. In terms of CPN, SFCA allows several winners to be active and therefore, naturally works in an nterpolation *mode. On the contrary, in the single-winner mode the CPN works in a hard matching manner by setting one and only one output signal of Kohonen unit to be 1, i.e.* $\zeta^j \in \{0, 1\}$. It is evident that the *interpolation* mode in CPN as suggested by Hecht-Nielsen bears some resemblance to the approximate reasoning process undertaken by SFCA. A significant implication of this analogy is that the matching degree interpretation and associated algorithm in SFCA provide an alternative and immediate solution to the split problem encountered in applying the *interpolation* mode for CPN, a problem which is identified by Hecht-Nielsen and discussed in the end of the previous subsection.

## 9.4 Self-construction of the rule-base

So far the knowledge representation and reasoning problems have been solved by structural mapping and the associated forward algorithm provided that the rule-base is available. It is noted that no training process is needed under this assumption because of the close correspondence which exists between the rule-base and the net structure. It would be apparently insignificant if we merely made a simple structural equivalence without taking full advantage of the adaptive property of neural networks. In other words, this localized knowledge representation paradigm would be more attractive and more useful only by removing the assumption of availability of the rule-base, thereby representing an extremely challenging problem which is referred as to knowledge acquisition and is essential for the creation of a knowledge-based system. To this end, the adaptive property of the network must be fully exploited to provide self-organizing and self-learning of the required rule-base directly from the controlled environment. In terms of CPN, this means that the number of units in the Kohonen layer must be self-organized and the associated weights with each unit must be self-learned. In the following, we first describe the standard training algorithm of the CPN, then present a modified scheme capable of carrying out the above task.

### 9.4.1 CPN training algorithm

Basically, the CPN training algorithm is a supervised training process by which the weights of the network are determined by exposing it to a set of paired training samples $(\mu^s, \nu^s)$. The algorithm consists of two parts: a Kohonen scheme which is unsupervised in nature and is used to learn $\omega^j$, and a Grossberg scheme which is truly supervised and is used to learn $\pi^j$.

More specifically, assuming that the number of Kohonen units is specified in advance and remains fixed during the training, the Kohonen algorithm can be described as follows. The training vector $\mu^s$ drawn randomly from $A$ is presented to the CPN, and the input layer distributes $\mu^s$ to each of the units of the Kohonen layer through connecting weight vectors $\omega^j$. Competition then takes place among all the Kohonen units so as to determine which unit wins the current competition by determining which unit is closest to the current $\nu^s$ in the sense of the defined metric. Following the competition, the output of the winning unit $J$ is set to 1, whereas outputs of all the other units are set to 0. The above two steps follow the same computing equations (9.10) and (9.11) as used during normal operation. The winner's weight vector $\omega^J$ is then updated by adding a time-varying scaled difference $\mu^s - \omega^J$ such that the modified $\omega^J$ is closer to $\mu^s$, whereas the weights of all the other units remain unchanged. More formally, the update law is given by

$$\omega^j(t) = \omega^j(t-1) + \alpha_t \cdot [\mu^s(t) - \omega^j(t-1)] \cdot \zeta^j \tag{9.14}$$

where $\zeta \in \{0, 1\}$ is determined by (9.10) and (9.11). $0 < \alpha_t < 1$ is a gain sequence decreasing monotonically with time in a linear or exponential manner. It is well known that if the coefficients $\alpha_t$ satisfy

$$\lim_{p \to \infty} \sum_{t=1}^{p} \alpha_t = +\infty \tag{9.15}$$

$$\lim_{p \to \infty} \sum_{t=1}^{p} \alpha_t^2 < \infty \tag{9.16}$$

then convergence can be guaranteed in the sense of mean squared performance measure. In what follows, a harmonic series $\alpha_t = 1 / t$ satisfying the above mentioned conditions will be used. The resultant weight vectors $\omega$ following the above training procedure can be considered to be approximately equi-probable with respect to $\mu$ vectors in a nearest-neighbour sense (Hecht-Nielsen 1990) and to be $N$ optimally quantized representatives of the input space $A$ in some metric sense.

Once the Kohonen layer has stabilized, $\omega^j$ can be frozen and the Grossberg layer begins to learn the desired output $\nu$ for each frozen weight vector $\omega^j$ by adjusting the weights connecting the Kohonen units to the Grossgerg units. More specifically, the update law at this layer is given by

$$\pi_k^j(t) = \pi_k^j(t-1) + \beta \cdot [-\pi_k^j(t-1) + \nu_k^s] \cdot \zeta^j \tag{9.17}$$

where $\pi_k^j$ is a weight from the $j$th Kohonen unit to the $k$th Grossberg (output) unit. $\beta$ is a constant update rate within the range [0,1]. $\nu_k^s$ is the $k$th component of the training sample $\nu^s$.

Since each time only the winning Kohonen unit $J$ produces the output $\zeta^J = 1$, equation (9.17) essentially modifies those weights connecting the $J$th unit to all the Grossberg units. By assertion made in the last section, weight vector $\pi^J$ is updated only. It has been observed that the learned $\pi^j$ in (9.17) are the exponential average of $\nu^s$ vectors associated with $\mu^s$ vectors which led the $j$th Kohonen unit to win, or in other words, those $\mu^s$ which are within the neighbourhood of $\omega^j$.

### 9.4.2 Self-organizing of the IF part of the rule-base

There are several difficulties in applying the CPN training algorithm described above to our case where real-time adaptation is required. However, the major problems lie in the fact that: (a) the number $N$ of the Kohonen units must be specified in advance; and (b) the correct or desired output at the Grossberg layer must be supplied. By assumption, the above required knowledge is unavailable and instead must be learned on-line as well as the learning of associated weight vectors. This section focuses on the problems relevant to the Kohonen layer or equivalently the IF part of the control rule, leaving the problems associated with the Grossberg layer or the THEN part to be dealt with in the next subsection.

Since the CPN network functionally plays the role of a controller, the distribution feature of the input data varies from case to case in an unknown manner depending largely on the characteristics of the controlled process, the form of command inputs and performance requirements. In these circumstances it is very difficult to specify the number of the cluster centroids, or equivalently of the control rules, especially if a high dimensional input of the CPN is involved. Therefore, it is necessary to develop a modified Kohonen algorithm so as to learn not only the weight vectors $\omega$ as the original algorithm does, but also the required number of the Kohonen units. Thus, we need a truly self-organizing learning algorithm and a dynamically variable CPN structure in respond to on-line incoming data.

Comparing the IF part of (9.9) in SFCA with (9.13) in CPN, there is a width vector $\Delta_u^j$ associated with the former. As discussed earlier, $\Delta_u^j$ can be visualized roughly as defining a neighbourhood for the $j$th rule centred at $M_u^j$. This viewpoint, together with the concept of relative distance (9.7), provides some insight into finding a solution for the problem. A vital idea is to associate with each existing Kohonen unit a valid radius and a local update gain such that each subregion represented by the weight vector $\omega$, and restricted by the associated radius, is treated as a completely localized region although the overlap along the boundaries between the adjacent regions is allowed. By assigning each Kohonen unit a predefined width vector $\Delta_u^j$, the winner $J$ not only has a minimum distance among all the existing units in regard to the current input $\mu$ as determined by (9.10), but also must satisfy the condition of $\mu$ falling into the winner's neighbourhood as designated by $\Delta_u^j$. Thus, if these two conditions are met, then unit $J$ is considered to be the winner and the associated weight vector is adjusted using (9.14) but with a local update gain. On the other hand, if these conditions are not satisfied with respect to all the units, it indicates that no existing unit is adequate to assign $\mu$ as its member and therefore, a new unit should be created. It is clear that, starting from an empty state, the Kohonen layer can be dynamically self-organized in terms of the number of units and weight vectors $\omega$ associated with each unit, thereby establishing the IF part of the rule-base.

Incorporating the above ideas into the standard Kohonen algorithm, we develop a modified algorithm and present it as follows, where $l$, $t$, and $T$ denote the iteration number, the sampling instant, and maximum sampling time respectively. In addition,

$\alpha^j$ is the local gain controlling the speed of the adaptive process for $\omega^j$, and is inversely proportional to the active frequency $n^j$ of the $j$th unit up to the present time instant, $N^l(t)$ stands for the number of the Kohonen units at the $l$th iteration and at time $t$, and it is assumed that all $\delta_l^j$ are identical being denoted by $\delta$.

*Modified Kohonen algorithm.*

1. Initialization. $\omega_1^1(0) = u(0)$; $n_1^1(0) = 1$; $\alpha_1^1(0) = 1 / n_1^1(0)$; $N_1(0) = 1$.
2. At the $l$th iteration, do the following at each sampling time $t$:
   (a) Find the unit $J$ which has the minimum distance to the current $\mu$ by

$$D(\omega_l^J, \mu) = \|\omega_l^J(t) - \mu(t)\| = \min_{j=1,\, N} \|\omega_l^j(t) - \mu(t)\| \tag{9.18}$$

   (b) Determine the winner using the following rule:

$$\begin{cases} \text{if} \quad D(\omega_l^J, \mu) \le \delta & \rightarrow \ J \textit{ is the winner} \\ \text{if} \quad D(\omega_l^J, \mu) > \delta & \rightarrow \ \textit{create a new unit} \end{cases} \tag{9.19}$$

   (c) Modify or initialize parameters:
   If $J$ is the winner:

$$\begin{cases} n_l^J(t) = n_l^J(t-1) + 1; \quad \alpha_l^J(t) = \dfrac{1}{n_l^J(t)} \\ \omega_l^J(t) = \omega_l^J(t-1) + \alpha_l^J \cdot [\mu(t) - \omega_l^J(t-1)] \\ N_l(t) = N_l(t-1) \\ \zeta^J = 1 \end{cases} \tag{9.20}$$

   If a new unit is created:

$$\begin{cases} N_l(t) = N_l(t-1) + 1 \\ \omega_l^{N_l}(t) = \mu(t) \\ n_l^{N_l}(t) = 1 \\ \zeta^{N_l} = 1 \end{cases} \tag{9.21}$$

3. After the $l$th iteration, remove all inactive units and reinitialize all active units:

$$\begin{cases} N_{l+1}(0) = N_l(T) - N_l^{inactive} \\ \omega_{l+1}^j(0) = \omega_l^j(T) \\ \alpha_{l+1}^j(0) = \alpha_l^j(T) \\ j = 1, 2, \cdots, N_{l+1}(0) \end{cases} \tag{9.22}$$

Compared with the standard Kohonen algorithm, the proposed algorithm above possesses two notable features. The first is that the input space is partitioned in a soft or fuzzy manner in the sense that the each Kohonen unit occupies a hyper-region, part of which is shared by other existing neighbouring clusters. The second feature is that the partition of the input space is a dynamical process altering the structure of CPN continuously along with both temporal and spatial dimensions.

### 9.4.3 Self-learning the THEN part of the rule-base

As mentioned previously, the major difficulty in deriving the weight vectors $\pi$ connecting the Kohonen layer to the Grossberg layer, or equivalently the THEN parts of the rules, stems from the unavailability of the teacher signals $v^j$ guiding the supervised training. More generally, lacking a teacher is a common obstacle facing any kind of neurocontroller and can be regarded as a bottleneck. Although considerable effort has been recently devoted to constructing the teacher signals explicitly or implicitly for neurocontrollers with various net topologies (e.g. Barto 1990), it is far from being completely and satisfactorily solved. Here, we propose a simple but efficient scheme capable of carrying out the task of training $\pi$. The approach comprises essentially two steps. First, the teacher signals are constructed explicitly at the beginning of each iteration by an iterative learning approach. Then the derived signals are supplied to the Grossberg layer of the CPN so as to adjust the $\pi$ vectors using the standard algorithm (9.17).

Figure 9.2 shows a block diagram of the learning system consisting of a reference model and a learning mechanism. The reference model is designed to specify what the process responses $y_p$ should be when both the model and the process are subject to the same command signal $r$. For the sake of simplicity, we again adopt the noninteracting model with second-order linear transfer functions.

Denoting the learning error by $e_L$, defined as the difference between the output $y_d$ of the reference model and the output $y_p$ of the process, the overall goal of the learning system is to force the learning error $e_L(\mathrm{t})$ asymptotically to zero or to a predefined tolerant region $\varepsilon$ within a time interval of interest $[0,T]$ by repeatedly operating the system. More specifically, we require that $\|e_{L,l}(t)\| \to 0$ or $\|e_{L,l}(t)\| < \varepsilon$ uniformly in $t \in [0,T]$ as $l \to \infty$, where $l$ denotes the iteration number, $\varepsilon > 0$ is a predefined error tolerance, and $\|\cdot\|$ stands for norm. It is clear that whenever convergence occurs, the corresponding control action at that iteration is regarded as the desired control $\gamma_d$ with which the corresponding response $y_p$ of the controlled process will be close enough to $y_d$.

By taking the process time delay into account, the learning law is given by

$$\gamma^{l+1}(t) = \gamma^{l}(t) + P_L \cdot e_{L,l}(t+\lambda) + Q_L \cdot c_{L,l}(t+\lambda) \tag{9.23}$$

where $\gamma^{l+1}, \gamma^{l} \in R^m$ are on-line learning teacher vector-valued functions at the $(l+1)$th and the $l$th iterations respectively, $e_{L,l}$, $c_{L,l} \in R^m$ are learning error and

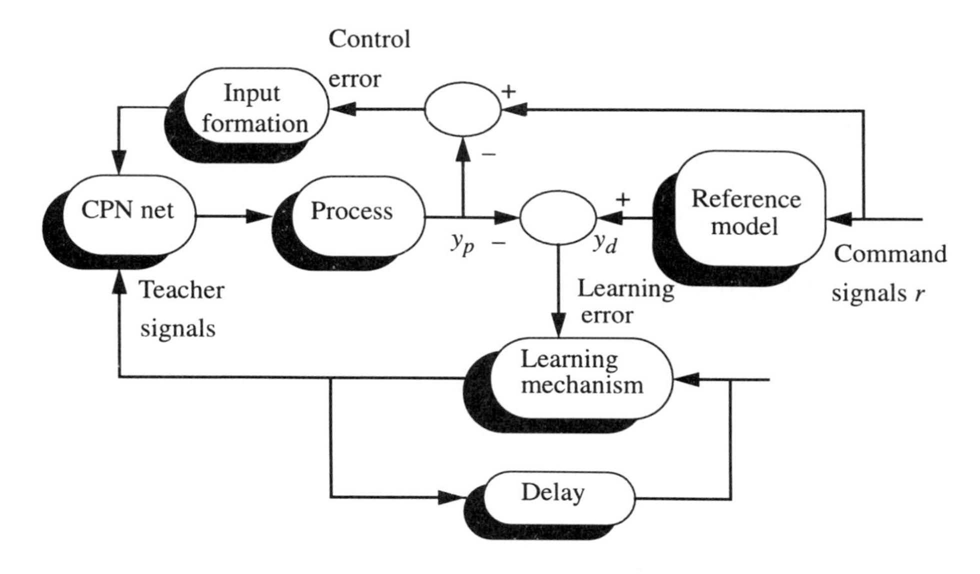


**Figure 9.2** A diagram for on-learning teacher signals

change of learning error defined by $c_{L,l}(t) = e_{L,l}(t + 1) - e_{L,l}(t)$, $\lambda$ is an estimated time advance corresponding to the time delay of the process, and $P_L, Q_L \in R^{m \times m}$ are constant learning gain matrices.

A variety of special algorithms can be obtained with different selections of learning gain matrices. For example, the simplest version is when $P_L$ is diagonal whereas $Q_L = 0$. In this case, the rate of change of the error is not considered, nor are the loop interacting effects. The learning action in each loop is totally dependent on its own error. Learning gain matrices $P_L$ and $Q_L$ should be chosen carefully. As would happen in any other learning system, too big or too small gain values will typically lead to either divergence or too slow convergence speed. Thus, a trade-off between speed and convergence should be made. In practice, this selection is not a very hard task because the learning system is relatively robust with respect to the values of the gains.

It is noted that the control update at the $(l+1)$th iteration is entirely based on the information derived at the $l$th iteration, i.e. the previous control action and the resulting performance measured by the error and its derivative. Accordingly, all the information at the $l$th iteration must be stored, and the all $\gamma(t) \in [0,T]$ at the $(l+1)$th iteration are adjusted simultaneously. Thus, at the beganing of each iteration, $\gamma$ is available at each time instant $t$ and can be used to guide the weight vector adjustment at the Grossberg layer by

$$\pi_k^j(t) = \pi_k^j(t-1) + \beta \cdot [-\pi_k^j(t-1) + \gamma_k^l(t)] \cdot \zeta^j \tag{9.24}$$

The above equation is the same as equation (9.17) except that here the desired signal $v^s$ is replaced by $\gamma$. Assume that, by appropriately choosing $P_L$ and $Q_L$, the desired output $y_d$ can be approached asymptotically with a learned control sequence $\gamma_d$. The $\gamma_d$ can then be embedded into $\pi^j$ vectors by equation (9.24) with a suitably chosen $\beta$. Thus, with increasing iterations, the THEN parts of the rules are gradually learned along with the learning process of $\gamma$.

## 9.5 Simulation results

The simulation studies were aimed at demonstrating the feasibility of the proposed scheme and examining its self-organizing and self-learning behaviours when applied to the multivariable blood pressure control problem as given in Appendix II. More specifically, we investigated the adaptive ability and convergence property of the system during the learning mode and the generalizing property in the reasoning mode. Throughout this work, the EC type controller defined in Section 9.2 was employed and the following values were used. The time interval of interest contains 100 sampling points with the sampling time being 30 s. Set-point changes for *CO* and *MAP* were set to be 20 ml/s and −10 mmHg changing from nominal values of 100 ml/s and 120 mmHg respectively. The maximum iteration number was 10. The other parameters used will be given accordingly.

1. Adaptive ability
   By adaptive ability, here we mean the capability of the system to deal with various situations with respect to controller environments, in particular to variations of the process parameters and desired performance requirements, and to the noise measurements, while keeping the controller parameters *fixed*. In this case, learning gain matrices $P_L$ = diag{0.05, −0.05} and $Q_L = 0$, Grossberg update rate $\beta$ = 0.5, and Euclidean distance with the valid radius $\delta = 0.1$ were used. To measure the performance, a normalized squared sum of errors (NSSE) has been defined by

$$NSSE^l = \frac{SSE^l}{SSE_{\max}} \tag{9.25}$$

   where

$$SSE^l = \sum_{i=1}^{100}\left[(y_{d,co}(i) - y_{p,co}{}^l(i))^2 + (y_{d,map}(i) - y_{p,map}{}^l(i))^2\right] \tag{9.26}$$

   is the squared sum of error at the $l$th iteration and $SSE_{\max}$ is the largest $SSE^l$ chosen from all the $SSE^l$ produced by using different parameters within the same group at all the iterations from $l = 1$ to $l = 10$. Figure 9.3 shows the results of NSSE versus the iteration number when different performance requirements were demanded by specifying the parameters in the reference model as follows:

   (a) $\xi_{co} = 0.8 \quad t_{s,co} = 240$ s $\quad \xi_{map} = 0.7 \quad t_{s,map} = 240$ s

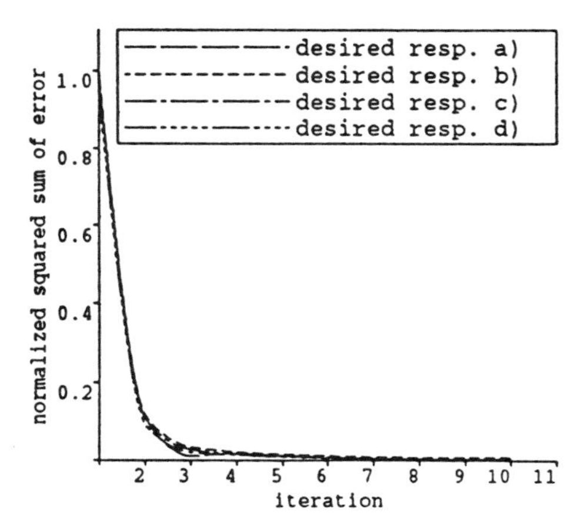


**Figure 9.3** Adaptive ability to desired responses

(b) $\xi_{co} = 0.7$ $t_{s,co} = 240$ s $\xi_{map} = 0.8$ $t_{s,map} = 240$ s
(c) $\xi_{co} = 0.8$ $t_{s,co} = 240$ s $\xi_{map} = 0.8$ $t_{s,map} = 240$ s
(d) $\xi_{co} = 0.9$ $t_{s,co} = 400$ s $\xi_{map} = 0.7$ $t_{s,map} = 240$ s

It can be seen that the system is capable of handling the various desired responses using a set of fixed controller parameters. The NSSE quickly approach small and stable values indicating a fast learning process. It can be seen that a relatively bigger NSSE was observed for case (b). In fact, the actual process responses are quite acceptable even for this seemingly worst case as indicated in Figure 9.4.

By altering the process parameters in the model given in Section 1.2.2 (Chapter 1), a set of NSSE were obtained and depicted in Figure 9.5, where three cases are displayed with parameters variations of 10% from their nominal values. Although slightly different convergence behaviour with different cases was observed, the stable and small NSSE were achieved after a few iterations in all cases.

To examine the learning ability in a noise-contaminated environment, we carried out a simulation where the outputs of the process were corrupted by random noise with a uniform distribution, and therefore, both controller inputs and adaptive inputs were contaminated by noise. Figure 9.6 shows the outputs of the process after 10 iterations with a noise amplitude of 10% at the set-points. It can be seen that the responses are satisfactory.

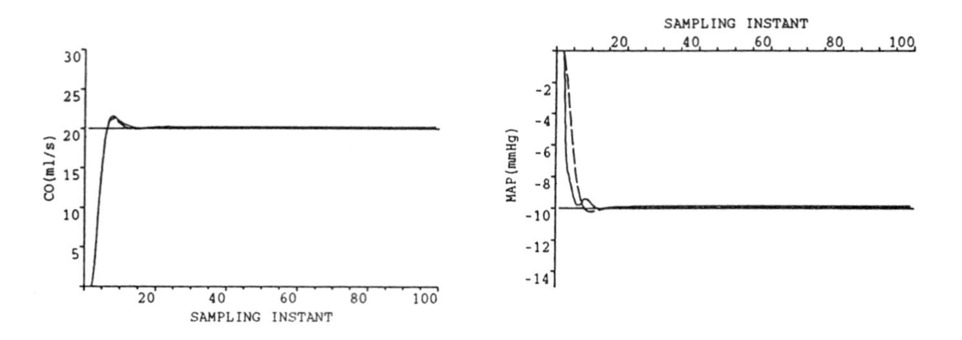


**Figure 9.4** Outputs of the process corresponding to case (b) of Figure 9.3

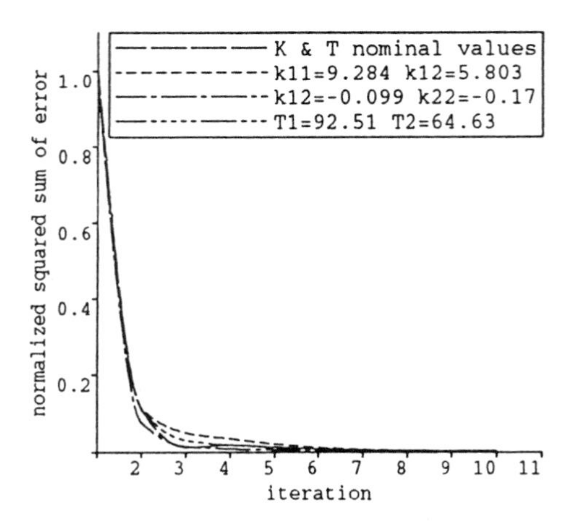


**Figure 9.5** Adaptive ability to process parameters

2. Convergence property
   By keeping the process parameters and the reference model fixed, the convergence properties of the system were investigated with respect to various controller parameters in terms of how these parameters influence the convergence process measured by NSSE.
   *Learning matrix $P_L$. The values of the learning matrix $P_L$* have a great effect on the learning process. Bigger values will speed up the convergence process but at the risk of divergence. In contrast, smaller values will slow down the learning

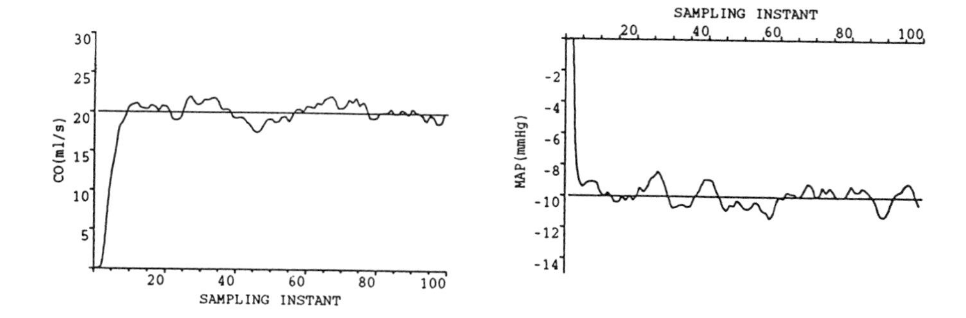


**Figure 9.6** Outputs of the process at the tenth iteration under noise measurements

process in, sometimes, an unacceptable manner. Figure 9.7 gives the results with $P_L$ = diag{0.05, –0.05}, diag{0.1, –0.1}, and diag{0.03, –0.03} respectively. As would be expected, the results agree well with the above remarks.

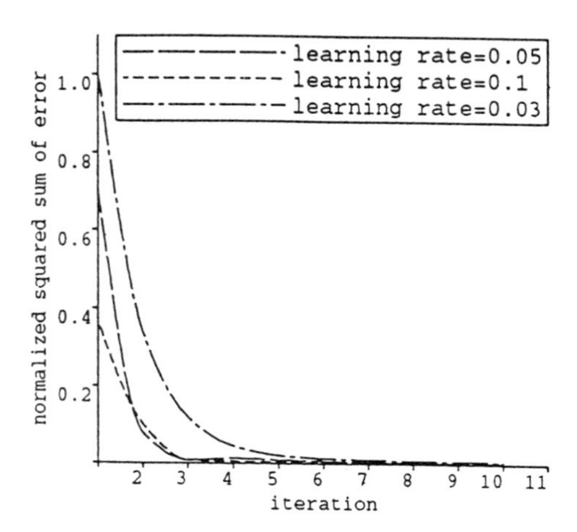


**Figure 9.7** Convergence property with learning rates

*Grossberg learning rate* $\beta$. *As mentioned earlier,* $\beta$ plays the role of controlling the update of $\pi$ in an exponential manner. The bigger it is, the more important the current teacher signal $\gamma$ is, and therefore the more quickly the previous learned $\pi$ is forgotten. Figure 9.8 shows the results with $\beta$ = 0.5, 0.1, and 0.9 respectively. It appears that too big or too small $\beta$ value will generally result in a relatively slower convergence process, indicating that values around 0.5 are good choices. However, the choice is not very sensitive.

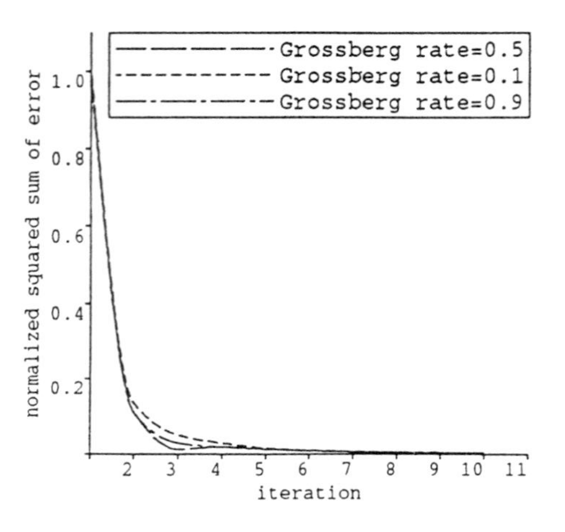


**Figure 9.8** Convergence property with Grossberg rate

*Valid radius* $\delta$. $\delta$ is mainly used to control the self-organizing process in the CPN. It has a direct effect on the number of Kohonen units or control rules. The bigger $\delta$ is, the more units the CPN creates, and therefore the more rules in the rule-base. By choosing $\delta$ = 0.01, 0.1, and 1.0 respectively, we obtained respective rule-bases with rule numbers being 22, 14, and 8. However, as expected, the convergence is not very sensitive to this parameter as indicated in Figure 9.9.

*Distance metrics.* Finally, we examined the effect of the distance metrics used in the self-organizing process at the Kohonen layer. Because different distance metrics define different shapes of the neighbourhood of each rule, it is expected that they have some effect on the number of created rules and less effect on the control or learning performances. We carried out comparative studies by using three frequently used metrics: Euclidean, Hamming, and Maximum distances $D_E$, $D_H$, and $D_M$. Three almost identical convergence processes, as shown in Figure 9.10, verified our expectation. However, it is interesting to note that the rule numbers produced by these metrics are 14, 16, and 13 respectively. These results

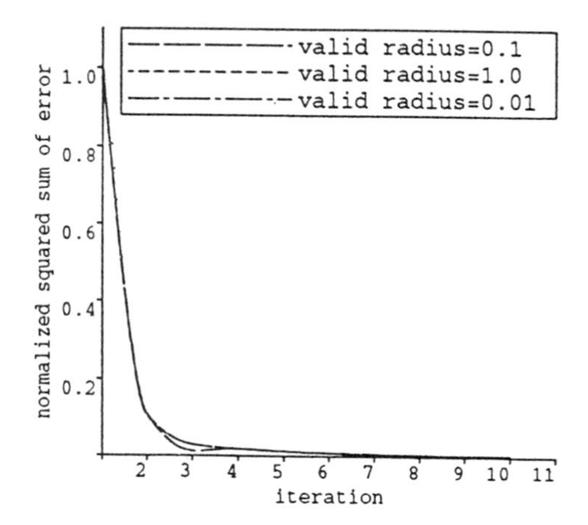


**Figure 9.9** Convergence property with valid radii

can be explained by the fact that, with the same valid radius δ, the neighbourhood defined by $D_M$ has the biggest "volume", therefore the smallest rule number 13; the one defined by $D_H$ has the smallest "volume", therefore the highest rule number 16, leaving the defined "volume" and the produced rule number by $D_E$ in the middle. Compared with the case of the valid radius, the different metrics have much less influence on controlling the rule number.

3. Approximate reasoning property
   The proposed CPN-based controller can work in two modes: real-time control with or without learning. In the latter case, it is assumed that an appropriate CPN structure with corresponding weights has been obtained during the learning stage. With this learned CPN or rule-base, it is expected that the controller can perform similar tasks in an acceptable manner by replacing a winner-take-all competitive scheme with a soft matching cooperative strategy. This requires the CPN to have some generalizing or interpolative capability or equivalently robustness with respect to some new situations. Table 9.1 gives the performance indices *ISE* and *ITAE* obtained at the tenth iteration during the learning mode (the first row) and those obtained using the learned CPN (14 rules) with the approximate reasoning scheme (the second row). The table indicates that the performances given by the latter are better than those given by the former. This may stem from the utilization of the interpolative reasoning. It is worth pointing out that for a safe use of the learned CPN, a bigger valid radius δ must be used to ensure the coverage of the whole possible input space by the learned rules. In our case, we assigned δ =

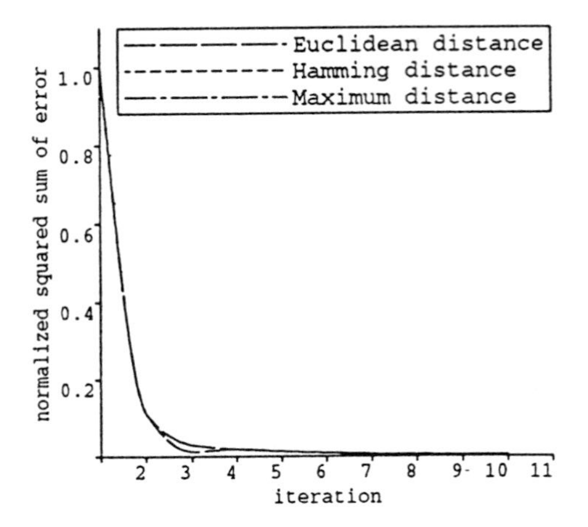


**Figure 9.10** Convergence property with distance metrics

0.6 instead of 0.1 which was used during the learning mode.

By using a learned CPN with 14 rules, we investigated the robustness property to variations of process gains, time constants, and a noise-contaminated environment.

**Table 9.1** Performance indices for learning and learned modes

| | $ISE_{co}$ | $ITAE_{co}$ | $ISE_{map}$ | $ITAE_{map}$ |
|---|---|---|---|---|
| Learning | 16.08 | 126.70 | 3.16 | 54.82 |
| Learned | 15.82 | 70.80 | 3.10 | 34.93 |

Figure 9.11 shows the results in terms of performance indices *Relative ITAE* (Figure 9.11(a)) and *Normalized ISE* (Figure 9.11(b)) versus the varying parameters, where

$$Relative\ ISE = \frac{|ISE^* - ISE|}{ISE^*} \tag{9.27}$$

$$Normalized\ ITAE = \frac{|ITAE^* - ITAE|}{Max|ITAE^* - ITAE|} \tag{9.28}$$

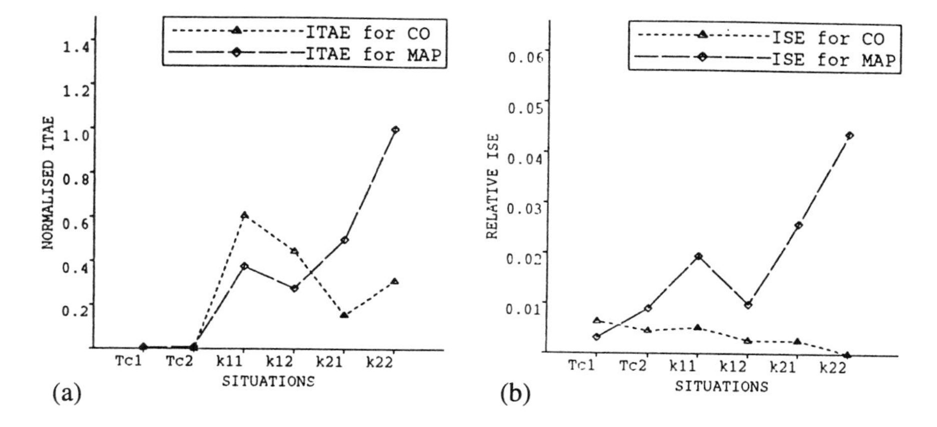


**Figure 9.11** Robustness of learned CPN controller

where $ISE^*(ITAE^*)$ and $ISE(ITAE)$ are the performance indices obtained with nominal and changed process parameter values (5% in this case) respectively. From Figure 9.11, we may conclude that (a) the system possesses good robustness as indicated by low *Relative ISE* values (Figure 9.11(b)); (b) the system is less sensitive to time constants than to process gains in the steady-state stage (Figure 9.11(a)) ; (c) in most cases, the *MAP* is more sensitive to the variation of parameters than *CO*. Finally, Figure 9.12 shows the output responses when the measured process outputs were corrupted by random noise with an amplitude of 10% at the set-points. It can be seen the system works satisfactorily in such a noise-contaminated environment.

## 9.6 Summary

We have introduced a unified framework for constructing automatically a control rule-base constrained by prespecified control performance requirements. The underlying principles of the approach rely on a combination of ideas from various fields. In particular, SFCA provides a very efficient interpolative look-up table paradigm suitable for use with multivariable numerical environments, CPN offers some insight into the idea of how a fixed-structure network can be developed adaptively into a multidimensional look-up table. Also, the concepts of relative distance in SFCA and the learning strategy in the learning control make it possible to extend the original

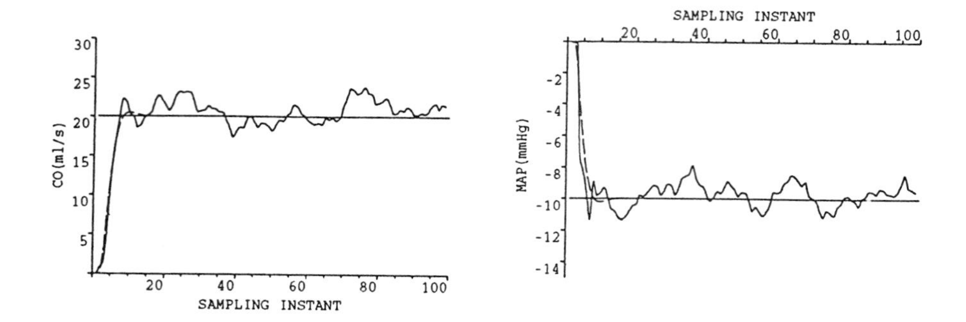


**Figure 9.12** Outputs of the process using learned CPN under noise measurements

CPN training algorithms at both Kohonen and Grossberg layers into a highly self-organized and completely unsupervised algorithm. The approach reported here outperforms the existing fuzzy control algorithms in many aspects. In particular, we claim that the approach is very generic in the sense that, in principle, control rule-bases of arbitrary dimensions can be constructed automatically while satisfying chosen desired, but physically achievable, performance requirements. The learning process involved is extremely fast due to the simple network topology and the efficient learning algorithms. Another point is that very little prior knowledge about the controlled process is required in the implementation of the algorithm but at the price of iterative operation, a price which must be paid regardless of whether using human or machine learning. Extensive simulations on a problem of multivariable blood pressure control have demonstrated the learning, adaptive, and approximate reasoning capabilities of the system. It has been found that the proposed approach is extremely efficient in terms of simple topological structure, fast learning speed, and good robustness properties.

CHAPTER 10

# Fuzzified CMAC and RBF network-based self-learning controllers

---

*A class of fuzzified localization networks is introduced with the associated scheme of active learning from controlled environments.*

---

## 10.1 Introduction

As viable alternatives to the well-known back-propagation neural networks, localized neural networks have increasingly gained popularity in many neural network application areas. The Albus's Cerebellar Model Articulation Controller (CMAC) (Albus 1975a, b) and the radial basis function network (RBF) (Broomhead and Lowe 1988; Poggio and Girosi 1990) are two typical paradigms. Although these networks share some common characteristics in one way or another, they are, in fact, rooted from rather different disciplines, leading to diverse interpretations and algorithms. While the CMAC has been inspired by the neurophysiological theory of the cerebellum, the RBF network is rooted primarily from function approximation or hypersurface fitting techniques, having little relevance to biological aspects.

The first objective of this chapter is to treat these networks in a unified way by looking them from a fuzzy system viewpoint. In this view, the trained network is structurally regarded as the means for representing rule-like knowledge in which each hidden unit specifies an IF-THEN statement with associated weights acting as the values of IF and THEN parts respectively. The function of the network is thought of as inferencing required conclusions in response to novel situations by a kind of approximate or fuzzy reasoning scheme. The above interpretation not only provides a new and clear insight into the deeper understanding of various versions of this kind of network, but also offers a flexible and unique computational algorithm encompassing, to some extent, the forward algorithms involved in the above mentioned networks. By integrating fuzzy algorithms with the CMAC or RBF network, we present a hybrid network, which may be called the fuzzified CMAC

(FCMAC) or fuzzified RBF network (FBFN) or more generally the fuzzified localization network (FLN).

The second objective of the chapter is to show how the integrated network can be employed as a multivariable self-organizing and self-learning controller. We consider the learning as a process of automatic knowledge acquisition from control environments. In particular, we present a dynamic learning algorithm capable of self-organizing both net structure and associated weights in a real-time manner. The feature of the simple net structure and fast on-line learning ability make the proposed approach particularly feasible for real-time control applications in which the network functions as an adaptive rule-based controller.

In Section 10.2, the structures of the CMAC and RBF networks are described. Section 10.3 is devoted to connecting the CMAC and RBF to the simplified fuzzy control algorithm SFCA, leading to the fuzzification of the CMAC and RBF networks. The learning algorithms are given in Section 10.4. The simulation results are presented in Section 10.5.

## 10.2 Description of the CMAC and RBF

### 10.2.1 CMAC network

In an attempt to derive an efficient computational algorithm for use in manipulator control, Albus developed a mathematical model called the Cerebellar Model Articulation Controller (CMAC) (Albus 1975a, b) which is based on the neurophysiological theory of the cerebellum. Figure 10.1 shows the structure of a CMAC network. Functionally, a trained CMAC performs a multivariable function approximation in a generalized look-up table fashion. Structurally, it is equivalent to a network architecture with three layers. With increasing interest in neural networks, CMAC has gained more and more attention particularly from control engineering researchers, due primarily to its unique characteristics such as fast training speed and localized generalization. For example, the learning control system developed by Tolle (Tolle *et al.* (1992); Tolle and Ersu 1992) utilizes two CMAC networks (or associative memory systems) as essential components, one for modelling the process and the other for storing the control strategy. Miller *et al.* (1990) have successfully applied the CMAC technique to robot control systems in which the CMAC functions as an inverse model of the robotic dynamics. Harris *et al.* (1992) have studied the control application of the CMAC network.

Recently, some improvements in the original CMAC have been reported, particularly in the aspects of addressing algorithms, training strategies, and convergence analysis. For instance, by introducing B-spline receptive field functions into the CMAC computational architecture, Lane *et al.* (1992) proposed a high-order CMAC network capable of approximating both functions and function derivatives. Brown

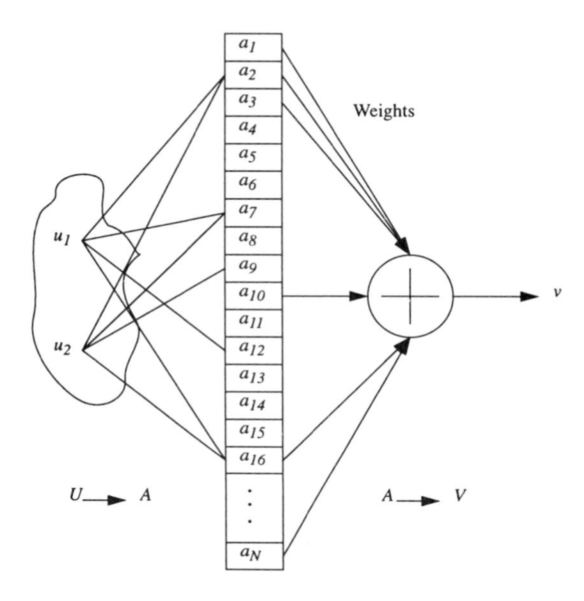


**Figure 10.1** A structure of CMAC

and Harris (1991) have made comparative studies on the connection between CMAC, B-splines and fuzzy logic control, pointing out some similarities between these schemes. The improvements to the design of CMAC systems have been proposed by Miller and co-workers and Parks and Militzer (Miller *et al.* 1990; An *et al.* 1991; Parks and Militzer 1991), including improved schemes for allocating weights to a given input vector and modifications to the design of receptive field shapes. Parks and Militzer (1992) presented five training algorithms with extensive comparative studies on the convergence property possessed by these algorithms. Theoretical results concerning the CMAC's learning convergence and approximation ability have been investigated based either on matrix equation theory (Ellison 1991; Parks and Militzer 1989; Wong and Sideris 1992) or on functional approximation theory (Cotter and Guillerm 1992). Very recently, Brown *et al.* (1993) investigated the interpolation capabilities of the binary CMAC.

One of the central issues concerning the design of a CMAC network is how to chose, in response to an input vector $u$, $N_L$ memory locations from $N$ $(N>N_L)$ locations where the function values are distributedly stored such that the summed value $v$ of $N_L$ weights is a reasonable approximation of the desired $v^*=f(u)$. This issue has been dealt with using a series of techniques consisting of discretizing, quantizing, coding, and hashing. Considerable complexity arises from using these procedures, particularly when a high-dimensional input space is involved. Once the addressing scheme is specified, the CMAC becomes a single-layered network and therefore, its data storage or training can be carried out in a simple way by any supervised training algorithm (e.g. Albus 1975b; Parks and Militzer 1992).

The CMAC is designed to represent approximately a multi-dimensional function by associating an input vector $u \in UR^n$ with a corresponding function vector $v \in VR^m$, where $U$ is usually assumed to be discrete and finite. As shown in Figure 10.1, the CMAC has a similar structure to a three-layered network with association cells playing the role of hidden-layer units. Mathematically, CMAC may be described as consisting of a series of mappings: $U \rightarrow A \rightarrow V$, where $A$ is an N-dimensional cell space.

A fixed mapping $U \rightarrow A$ transforms each $u \in U$ into an $N$-dimensional binary associate vector $a(u)$ in which only $N_L$ elements have the values of 1, where $N_L < N$ is referred to as the generalization width. In other words, each $u$ activates precisely $N_L$ association cells or geometrically each $u$ is associated with a neighbourhood in which $N_L$ association cells are include. An important property of the CMAC is local generalization derived from the fact that nearby input vectors $u_i$ and $u_j$ have some overlapping neighbourhood and therefore share some common association cells. The degree to which the neighbourhoods of $u_i$ and $u_j$ are overlapping depends on the Hamming distance $H_{ij}$ of $u_i$ and $u_j$. If $H_{ij}$ is small, the intersection of $u_i$ and $u_j$ should be large and vice versa. At some values of $H_{ij}$ greater than $N_L$, the intersection becomes null indicating that no generalization occurs.

According to the above principle, Albus developed a mapping algorithm consisting of two sequential mappings: $U \rightarrow M \rightarrow A$ which performs a content-addressing task. The $n$ components of $u$ are first mapped into $n$ $N_L$-dimensional vectors and these vectors are then concatenated into a binary association vector $a$ with only $N_L$ elements being 1. Albus also suggested an approach to reducing further the $N$ association cells into a much smaller set by a hashing code. Thus, the above mapping algorithm involves a series of procedures such as discretizing, quantizing, coding, and possibly hashing. It is evident that the computation and implementation complexity will increase dramatically with an increase of dimension in the input space, although the complexity may be reduced by including the hashing procedure.

The $A \rightarrow V$ mapping is simply a procedure of summing the weights of the association cells excited by the input vector $u$ to produce the output. More specifically, each component $v_k$ is given by

$$v_k = \sum_{j=1}^{N} a^j(u)\pi_k^j \tag{10.1}$$

where $\pi_k^j$ denotes the weight connecting the $j$th association cell to the $k$th output. Notice that only $N_L$ association cells contribute to the output.

### 10.2.2 RBF network

Recently Radial Basis Function (RBF) networks have been applied to many areas such as pattern recognition, signal processing, system modeling and control. Instead of being inspired from a biological background, the RBF network has its origin in function approximation and hypersurface fitting techniques, thereby having a well-established theoretical basis (Broomhead and Lowe 1988). Poggio (Poggio and Girosi 1990) provided a broad view of how the RBF is related to some classical approaches such as generalized splines, regularization theory, Parzen windows, and potential functions. A basic viewpoint concerning a three-layered RBF network is that it can represent a specific nonlinear function reasonably well by linearly combining a set of nonlinear and localized basis functions which span a space containing a class of functions to be approximated. It has proved by Park and Sandberg (1991, 1993) that under certain mild conditions on the basis functions RBF networks having one hidden layer are capable of universal approximation, thereby providing a sound foundation for use of RBF networks as approximators in various practical fields.

Simply stated, a RBF network is intended to approximate a continuous mapping f: $R^n \rightarrow R^m$ by performing a nonlinear transformation at the hidden layer and subsequently a linear combination at the output layer. More specifically, this mapping is described by

$$\hat{f}_k(u) = \sum_{j=1}^{N} \pi_k^j \cdot \phi^j(\|u - \omega^j\|) \tag{10.2}$$

where $N$ is the number of the hidden units, $u \in R^n$ is an input vector, $\omega^j \in R^n$ is the centre of the $j$th hidden unit and can be regarded as a weight vector from the input layer to the $j$th hidden unit, $\phi^j$ is the $j$th radial basis function or response function, and $\pi_k^j$ is the weight from the $j$th hidden unit to the $k$th output unit.

Although there exist many possibilities for the choice of $\phi^j$, as observed by Moody and Darken (1989), Gaussian type functions given by

$$\phi^j(u) = \exp[-\frac{\|u - \omega^j\|^2}{(\sigma^j)^2}] \tag{10.3}$$

offer a desirable property making the hidden units to be locally tuned, where the locality of the $\phi^j$ is controlled by $\sigma^j$. Thus, each hidden unit is associated with a centre vector $\omega^j$ and a width $\sigma^j$. As pointed out by Moody, the locality of the unit response functions is a vital factor for attaining fast learning speeds. This is because, for any given input, only a small fraction of the hidden units close enough to the input will be excited and therefore only those weights associated with the activating units will be evaluated and trained. By noting the factorable property of the Gaussian function, Poggio and Girosi (1990) discussed some intriguing analogies of the RBF to its neurobiological counterpart. Moody and Darken (1989) also proposed a normalized algorithm given by

$$\hat{f}_k(u) = \frac{\sum_{j=1}^{N} \pi_k^j \cdot \phi^j(\|u-\omega^j\|)}{\sum_{j=1}^{N} \phi^j(\|u-\omega^j\|)} \tag{10.4}$$

## 10.3 Connecting the CMAC and RBF to the SFCA

### 10.3.1 SFCA: A general form

The simplified fuzzy control algorithm (SFCA) described in Section 9.2 is, in fact, a special case of a class of more general algorithms consisting of two stages, pattern matching and weighted averaging as illustrated in Figure 10.2.

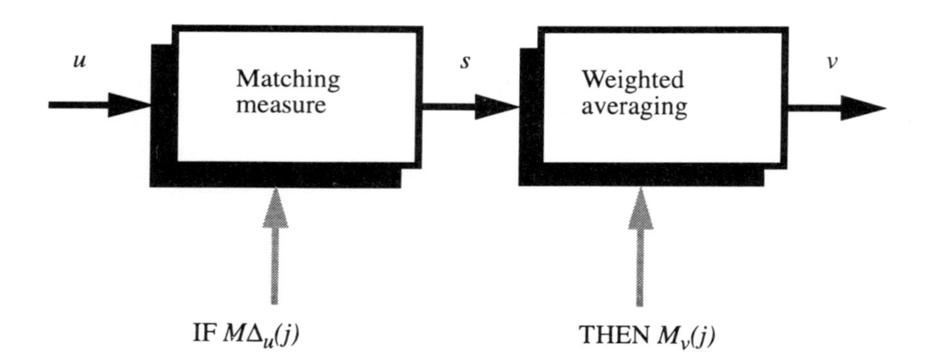


**Figure 10.2** An illustration of reasoning algorithm

To be more specific, we reformulate the problem as follows using the same notations as used in Section 9.2:

- Given $P$ simplified rules, each of which has the form,

$$IF \quad M\Delta_{u(j)} \quad THEN \quad M_v^j \tag{10.5}$$

where $M\Delta_u(j) = (M_u^j, \Delta_u^j)$ and

$$M_u^j = (M_{u,1}^j, M_{u,2}^j, \cdots, M_{u,n}^j)$$
$$\Delta_u^j = (\Delta_{u,1}^j, \Delta_{u,2}^j, \cdots, \Delta_{u,n}^j)$$

$$M_v^j = (M_{v,1}^j, M_{v,2}^j, \cdots, M_{v,n}^j)$$

- Given the current numeral input $u_0 = (u_{0,1}, u_{0,2}, \cdots, u_{0,n})$
- Find the numerical output $v = (v_1, v_2, \cdots, v_m)$

Then the reasoning algorithm for finding the $v_k$ is reformulated as follows:

- Calculate the *jth* matching degree $s^j \in [0,1]$ by

$$s^j = S(u_0, M\Delta_{u(j)}) \tag{10.6}$$

where $S(\cdot, \cdot)$ denotes the matching measure which can be defined in many ways. We give the following three possible definitions:

*Volume measure:*

$$s_v^j = \begin{cases} 1 - \dfrac{V_0}{V_R^j} & \text{if for } all\ i, |M_i^j - u_{0i}| \le \delta_i^j \\ 0 & otherwise \end{cases} \tag{10.7}$$

where $V_0^j = \prod_{i=1}^{n} |M_i^j - u_{0i}|$ is the volume of a hypercube taking every element $|M_i^j - u_{0i}|$ as its side, and $V_R^j = \prod_{i=1}^{n} \delta_i^j$ is the volume of the hypercube formed by the *jth* rule pattern.

*Distance measure:*

$$s_d^j = 1 - D_j(u_0, M\Delta_u(j)) \tag{10.8}$$

where $D_j \in [0,1]$ denotes the relative distance defined by equation (9.7).

*Possibility measure:*

$$s_p^j = \mathop{\Phi}_{i=1}^{n} poss(C_i / A_i^j) \tag{10.9}$$

where $\Phi$ stands for the *AND* connective defined usually either as a *product* or as a *minimum* operation. In our case, since $C_i$ are singletons $u_{0i}$, equation (10.9) reduces to

$$s_p^j = \mathop{\Phi}_{i=1}^{n} A_i^j(u_{0i}) \tag{10.10}$$

where $A_i^j(\cdot)$: $U_i \rightarrow [0,1]$ are membership functions corresponding to fuzzy subsets $A_i^j$. Two common definitions of $A_i^j(\cdot)$ are triangle and exponential forms given by

$$A_i^j(u_i) = \begin{cases} 1 - \dfrac{|M_{u,i}^j - u_i|}{\delta_i^j} & \text{if } |M_{u,i}^j - u_i| \leq \delta_i^j \\ 0 & \textit{otherwise} \end{cases} \tag{10.11}$$

and

$$A_i^j(u_i) = \exp\left(\frac{-|M_{u,i}^j - u_i|^2}{\delta^2}\right) \tag{10.12}$$

Four different $s_p^j$ can be obtained by combining equations (10.11) or (10.12) with *product* or *minimum* operators of *AND*.

- Calculating the output $v_k$ by

$$v_k = \sum_{j=1}^{P} \hat{s}^j \cdot M_{v,k}^j \tag{10.13}$$

where

$$\hat{s}^j = \frac{s^j}{\sum_{j=1}^{P} s^j} \tag{10.14}$$

The above computational procedures are schematically illustrated in Figure 10.3.

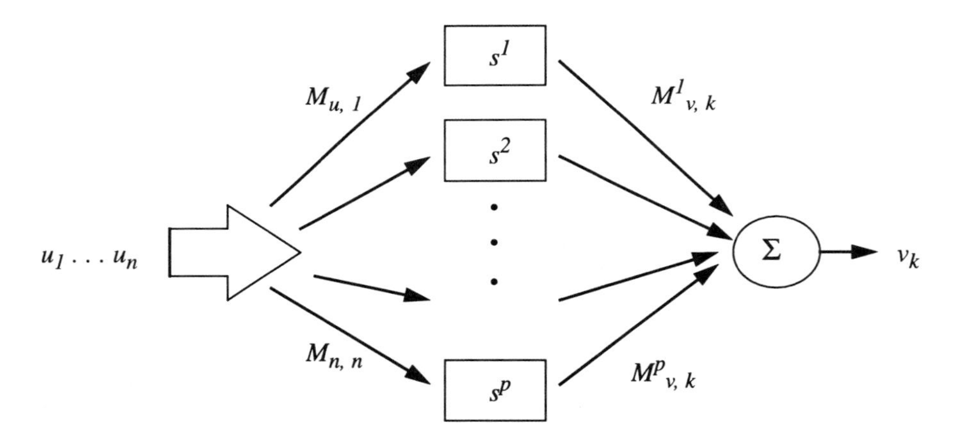


**Figure 10.3** Procedures of obtaining $v_k$ by the SFCA algorithm

### 10.3.2 FCMAC: Fuzzified CMAC

By carefully inspecting the SFCA and the CMAC, we have concluded that there exist some striking similarities between these two systems. Functionally, both of them perform a function approximation in an interpolative look-up table manner with an underlying principle of generalization and dichotomy: to produce similar outputs in response to similar input patterns and produce independent outputs to dissimilar input patterns. From the computational point of view, referring to Figures 10.1 and 10.3, mapping $U \to A$ corresponds to the calculation of the matching degree in the SFCA, and mapping $A \to V$ corresponds to the weighted averaging procedure given by equation (10.13). While the latter similarity is apparent by comparing equation (10.13) with equation (10.1), where $a^j$, $\pi_k^j$, and $N$ correspond to $s^j$, $M_{v,k}^j$, and $P$, the former equivalence can be made clearer as follows.

Instead of saying that each input $u$ is associated with a neighbourhood specified by $N_L$, we can equally consider that each association cell $\Psi^j$ is associated with a neighbourhood centred at, say $\omega^j \in U$ referred to as the reference vector, with the width controlled by $N_L$. If a current input $u$ is within the neighbourhood of $\Psi^j$, that cell is regarded as being active. In this view, the associate vector $a(u)$ can be derived by operating $N$ neighbourhood functions $\psi^j(\omega^j, u)$ with respect to $u$, where

$$\psi^j(\omega^j, \mu) = \begin{cases} 1 & \text{if } \mu \in \Psi^j \\ 0 & \text{otherwise} \end{cases} \tag{10.15}$$

that is, $a(\mu) = (\psi^1, \psi^2, \ldots, \psi^N)$. By appropriately selecting the $\omega^j$, $a(u)$ can be made to contain only $N_L$ 1's. Now it becomes evident that the associate vector $a$ is similar to the matching degree vector $s$ except that the former uses the crisp neighbourhood function, whereas the latter adopts the graded one. In fact, by letting $s^{*j} = 1$ for $s^j > 0$ and $s^{*j} = 0$ for other cases, the vector $a$ will be precisely equal to the vector $s^*$, indicating that the former is a special case of the latter. We notice that a natural measurement of whether $u$ belongs to $\Psi^j$ in equation (10.15) is to use some distance metric relevant to the generalization width $N_L$. In fact, Albus himself (1975a) discussed the question of how the overlapping of $u_i$ and $u_j$ is related to the Hamming distance of $u_i$ and $u_j$ although he did not formulate explicitly this concept into the mapping process $U \to A$.

Now we are ready to implement a fuzzified CMAC or FCMAC by replacing equation (10.1) with equation (10.13), where Albus's *coding algorithm-based content-addressing* technique for calculating the association vector $a$ is replaced by our *matching measure-based content-addressing* approach; or more specifically, the crisp neighbourhood function (10.15) is replaced by the matching measure (10.6). Several advantages can be identified by this replacement. The concept of the graded matching degree not only provides a clear interpretation for $U \to A$ mapping, but also offers a much simpler and more systematic computing mechanism than that proposed by Albus where some very complicated addressing techniques are utilized and further hashing code may be needed to reduce the storage. In addition, as noted by Moody and Lane *et al.* (1992), the graded neighbourhood functions overcome the problem of discontinuous responses over neighbourhood boundaries due to the crisp neighbourhood functions.

### 10.3.3 FBFN: generalized RBF networks

As in the case of CMAC, some intrinsic similarities between the RBF and the SFCA can also be found, although they have originated from two apparently independent fields: fuzzy logic theory and function approximation theory. From the knowledge representation viewpoint, the RBF network is essentially a net representation of IF-THEN rules. Each hidden unit reflects a rule: *IF* $\omega^j$ *THEN* $\pi^j$, where $\pi^j = (\pi_1^j, ..., \pi_m^j)$. The rationale behind fuzzy reasoning in the SFCA and interpolative approximation in the RBF seems to be the same: to create a similar action with respect to a similar situation or to produce a similar output in response to a similar input. By comparing equation (10.13) with equations (10.2) or (10.4), we see that the two systems have almost identical computational procedures i.e. matching degrees correspond to response function values and the resemblance of two approaches in the final combination step is evident. The following parameter correspondences are identified: $M\Delta_u(j)$ to $\omega^j$, $M_v^j$ to $\pi^j$, $\delta$ to $\sigma$, and $P$ to $N$.

Once a formal connection between these two paradigms is made, we can derive a hybrid system taking advantages of both. One of the possibilities is to generalize the RBF network, by fuzzifying it, into a class of more general networks, referred to as fuzzified basis function networks or FBFN for short. This can be done by simply replacing the radial basis function $\phi^j$ with matching degree $s^j$. It is easy to verify that the Gaussian basis function (10.3) is exactly recovered if the exponential membership function (10.12) with *product* operator for *AND* is used. While the $s_d$ and $s_p$ in the above maintain the *radial* property, the other selections of $s_p$ are no longer $n$-D radial, instead 1-D radial. However, the similarity measurement interpretation of the basis function reveals that the global radiality is in fact not necessarily a prerequisite. What is important is to sustain the locality of $\phi^j$ in such a way that the similarity measure is reasonable and meaningful. In this regard, $\phi^j$ can be made not only factorable but also synthesizable, in a logical sense, as a result of the fuzzy operator *AND* on $n$ independent $\phi_i^j$ by *product* or *minimum*. It is surprising to note that the factorized $\phi_i^j$ are nothing but membership functions! Thus, we are led to an extremely controversial issue relevant to the basis of fuzzy theory, i.e. the subjectivity of the membership function, one of the sources of objection to fuzzy theory. Nevertheless, this subjectivity may provide an alternative explanation of why the choice of $\phi$ is not crucial.

*Remark:* Theoretical investigations (e.g. Park and Sandberg 1991, 1993) have proved that the RBF network is a universal approximator, meaning that any continuous function defined on a compact set of its input space can be uniformly approximated arbitrarily well by an appropriate RBF network with a sufficient number of hidden units. Likewise, by noting the close analogy between SFCA and RBF we may conclude that the fuzzy system working at a numerical level is a universal approximate reasoner. Thus, without the need of a rigorous mathematical proof, we can assure the approximate capability possessed by fuzzy systems, this being a benefit obtained from the cross-fertilization between two approaches.

## 10.4 Self-construction of the fuzzified network-based controller

The previous section indicates that localized networks with three layers such as the CMAC and RBF can be fuzzified into more general networks that may be referred to as fuzzified localization networks (FLN); or conversely, the CMAC and RBF may be regarded as the special cases of the FLN. The term FLN will be used in the discussion that follows without explicit reference to the FCMAC or FBFN. It is assumed that the FLN has $n$ input units with input vector $u \in U \subset R^n$, $N$ hidden units with basis functions $\phi^j$, and $m$ output units with the output vector $v \subset R^m$. A hidden unit specifies a rule: *IF* $(\omega^j, \Delta_u^j)$ *THEN* $\pi^j$, where $\omega^j$ and $\pi^j$ are weight vectors associated with the *jth* hidden unit. We notice that it is not difficult to find the equivalent notations used in the SFCA, FCMAC, and FBFN corresponding to those used in the FLN.

As mentioned previously, our main interests lie in using a FBFN, or a FCMAC, or more generally, a FLN network as a direct feedback controller which is self-constructed without relying on the control rules provided by human experts or the teacher signals supplied from other external sources. Thus, an active learning scheme that learns directly from control environments is needed. In terms of the FLN, parameters $\omega^j$, $\Delta_u^j$, $\pi^j$, and $N$ must preferably be self-determined without knowing the teacher signals $v^*$. The algorithms developed for training of the traditional CMAC and RBF networks are not directly applicable to our purpose. Most of the algorithms involve learning the weight vector $\pi$ only, leaving the net structure parameter $N$ and the weight vector $\omega$ to be determined in advance. A notable exception is due to Moody and Darken (1989) who developed a hybrid learning scheme to learn $\omega$ unsupervisedly, $\Delta_u$ heuristically, and $\pi$ supervisedly. Since the FLN-based controller must be operated in real-time, there are some difficulties in applying Moody's method. In particular, it is hard to specify $N$ in advance due to the uncertain distribution of on-line incoming data, and more seriously, there are in general no teacher signals available to guide the learning of $\pi$. We approach the problem by using the same idea of constructing the CPN networks as used in the previous chapter but with some modifications so as to make the algorithm suitable for constructing the FLN network.

More specifically, the FLN-based controller is self-organized by a competitive learning scheme combined with a supervised training algorithm with the teacher signals being learned on-line as shown in Figure 10.4. Assuming that an identical width $\delta$ is used and prespecified on the basis of some knowledge about the input

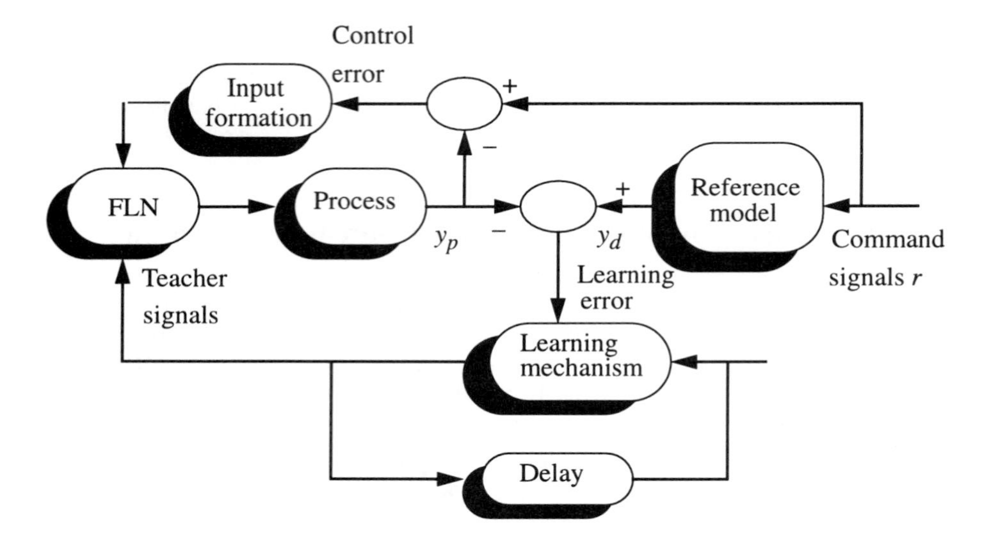


**Figure 10.4** FLN-based learning control system

space, the structure parameter $N$ and associated weight vectors $\omega$ are dynamically learned in response to the on-line incoming controller input $u$.

1. Calculate the $\phi$ vector by equation (10.6), where

$$\phi = (\phi^1(\omega^1, u), \phi^2(\omega^2, u), \ldots, \phi^{N(t)}(\omega^{N(t)}, u)) \tag{10.16}$$

2. Find the unit $J$ having the maximum response value:

$$\phi^J = \max_{j=1,N} \phi^j \tag{10.17}$$

3. Determine the winner using the following rule:

$$\begin{cases} \text{if } \phi^J \geq \phi_0 & \rightarrow J \textit{ is winner} \\ \text{if } \phi^J < \phi_0 & \rightarrow \textit{create a new unit} \end{cases} \tag{10.18}$$

   where $0 \leq \phi_0 < 1$ is a threshold controlling the unit number created.

4. Modify or initialize parameters:

   If $J$ is the winner:

$$n_h^J(t) = n_h^J(t-1) + 1; \quad \alpha^J(t) = \frac{1}{n_h^J(t)}; \quad N(t) = N(t-1)$$
$$\omega^J(t) = \omega^J(t-1) + \alpha^J \cdot [u(t) - \omega^J(t-1)] \tag{10.19}$$

where $n_h^J(t)$ is the active frequency of the *Jth* unit up to the current time instant.

If a new unit is created:

$$N(t) = N(t-1) + 1;\ \omega^{N(t)} = u(t);\ n_h^{N(t)} = 1;\ \phi^{N(t)} = 1 \qquad (10.20)$$

5. Output the $\phi$ vector.

It is worth noting that the above procedures are completed at each sample instant. After each iteration it is necessary to perform a check procedure to retain only those hidden units that have been active at least once during the latest iteration, and to discard all other inactive units.

Now the algorithm for learning $\pi$ vectors is given as follows. First, we notice that if the $k$th desired network output $v_k^*$ corresponding to the current FLN input $u$ is known, then the learning rule for $\pi_k^j$ can be derived easily following the standard instantaneous gradient procedure by finding the gradient of an instant squared error measure with respect to $\pi_k^j$ and noting equation (10.13), thus giving

$$\pi_k^j(t+1) = \pi_k^j(t) + \frac{\beta(v_k^* - v_k(u))\ \phi^j}{\sum_{j=1}^{N} \phi^j} \qquad (10.21)$$

where $0 < \beta < 1$ is a learning rate and $v_k(u)$ is the actual output of the FLN. As pointed out previously, however, the major difficulty in applying equation (10.21) lies in the unavailability of teacher signals $v_k^*$ for guiding the supervised training. Here the required teacher signals $v_k^*$ are explicitly constructed at the beginning of each iteration, as shown in Figure 10.4, by the following iterative learning algorithm:

$$v^{*l+1}(t) = v^{*l}(t) + p_L \cdot e_{L,l}(t+\lambda) + q_L \cdot c_{L,l}(t+\lambda) \qquad (10.22)$$

where $v^{*l}, v^{*l+1} \in R^m$ are on-line learning teacher vector-valued functions at the $l$th and the $(l+1)$th iterations respectively, $e_{L,l}$, $c_{L,l} \in R^m$ are learning error and change of learning error defined by $e_{L,l} = y_d - y_p$ and $c_{L,l}(t) = e_{L,l}(t+1) - e_{L,l}(t)$ respectively, $\lambda$ is an estimated time advance corresponding to the time delay of the process, and $p_L$, $q_L \in R^{m \times m}$ are constant learning gain matrices.

*Remark 1:* The learning algorithm given above differs from that described in Chapter 7 mainly in the following ways. Instead of using the minimum distance as the winner-selection criterion, here we employ the matching degree to determine the winner. Instead of using the winner-take-all exponential learning scheme for learning $\pi$, here normally more than one $\pi$ vector are updated, determined by the corresponding matching degree.

*Remark 2:* The update law of equation (10.21) is applied only when there is no new unit needed to be created at the time instant $t$ as determined by equation (10.18). However, if it is decided to create a new unit, then the response value associated with the new unit is set to be 1 and the corresponding $\pi$ vector is initialized to the current learned teacher vector $v^*$ which is available at the beginning of each

iteration.
*Remark 3:* Assuming that the $k$ desired output $v_k^*$ corresponding to the current CMAC input $u$ is known, Albus (1975b) developed a training algorithm spreading the output error evenly to a fixed $N_L$ association cell weights contributing to the present output $v_k(u)$, which is given by

$$\pi_k^j(t+1) = \pi_k^j(t) + \frac{\beta[(v_k^* - v_k(u))a^j]}{N_L} \tag{10.23}$$

Notice that only $N_L$ weights are adjusted by the same amount due to the fact that the association vector $a$ consists of exactly $N_L$ 1's. It is evident that the algorithm of equation (10.23) is essentially a special case of the gradient descent algorithm of equation (10.21).
*Remark 4:* Although many different learning algorithms can be derived from equation (10.22) with different selections of matrices $p_L$ and $q_L$, the simplest one is when $p_L$ is diagonal whereas $q_L$=0. In this case, the rate of change of the error is not considered, nor are the loop-interaction effects, and therefore, the learning action in each loop is totally dependent on its own error. This simple algorithm in fact works well in dealing with multivariable control problems as will be demonstrated in the next section.
*Remark 5:* The structural correspondence between the SFCA algorithm and the FLN enables the easy incorporation of prior control knowledge into the FLN-based controller. This can be done by expressing control knowledge as a set of rules with the form of equation (10.5) and using them as initial parameters of the FLN. It is expected that this would speed up the learning process if the knowledge provided by humans is correct. On the other hand, any incorrect or inconsistent knowledge will be automatically compensated via the learning.

## 10.5 Simulation results

We carried out a set of simulations on the problem of blood pressure control using the model given in Appendix II. The objectives of the simulation were to show the feasibility of the FLN-based control scheme when applied to this multivariable problem, to compare the performance with various response functions $\phi$, and to examine the self-construction behaviour of the system. Throughout the simulation, the controller was composed of six input and two output variables. The six input variables are two errors, two change-in-errors, and two sum-errors.

With the controller parameters set as follows: learning matrices $p_L = diag\,0.05,-0.05$ and $q_L = 0$, learning rate $\beta = 0.1$, threshold $\phi_0 = 0.1$, and width $\delta = 5$, we conducted the simulations with four different basis functions:

$$\phi_{tp}^j = \prod_1^6 [1 - \frac{|u - \omega^j|}{\sigma}] \tag{10.24}$$

$$\phi_{tm}^{j} = \bigwedge_{1}^{6} [1 - \frac{|u-\omega^{j}|}{\sigma}] \tag{10.25}$$

$$\phi_{ep}^{j} = \prod_{1}^{6} \exp[- \frac{|u-\omega^{j}|^{2}}{\sigma^{2}}] \tag{10.26}$$

$$\phi_{em}^{j} = \bigwedge_{1}^{6} \exp[- \frac{|u-\omega^{j}|^{2}}{\sigma^{2}}] \tag{10.27}$$

subject to step commands, where $\Pi$ and $\Lambda$ denote algebraic product and minimum operator respectively. Corresponding to $\phi_{tp}^{j}$, $\phi_{tm}^{j}$, $\phi_{ep}^{j}$, and $\phi_{em}^{j}$, the squared sum of learning errors $e_L$ (SSE) after 15 iterations were 60.42, 60.51, 60.76, and 60.77 and the number of created hidden units (rules) were 8, 8, 7, and 7 respectively. The results indicated that the control performance is not sensitive to the choice of the basis function and therefore it can be designed in various ways. This agrees well with our previous conclusion. This property may have some significant implications. For example, synthesizing the global function from 1-D functions with the minimum operation is much easier to implement using hardware than any others. In what follows, only $\phi_{ep}$ was adopted.

Figure 10.5 shows the output responses of the process after 15 iterations together with the desired responses indicated by dashed lines, where the measured process outputs were corrupted by random noise with an amplitude of 10% at the set-points.

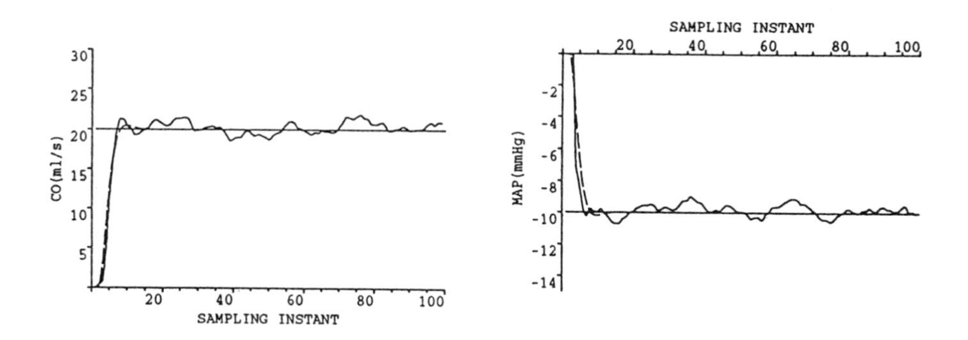


**Figure 10.5** Output responses of the process under noise measurements

It can be seen that the process outputs follow the desired responses satisfactorily although both controller inputs and learning inputs were contaminated by noise.

To investigate the adaptive ability of the system to various situations, we conducted the following simulations using the above fixed controller parameters. Figure 10.6 shows the results when four different desired responses (obtained by changing some parameters in the reference model) were required, whereas Figure 10.7 gives the results with different process parameters being changed by 10% from their nominal values.

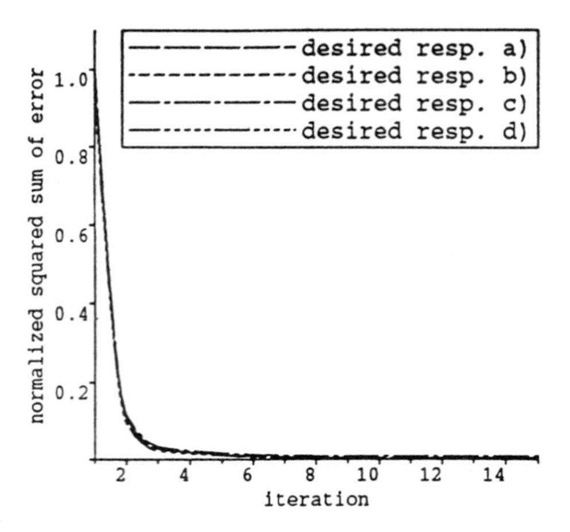


**Figure 10.6** Adaptive ability to the desired responses

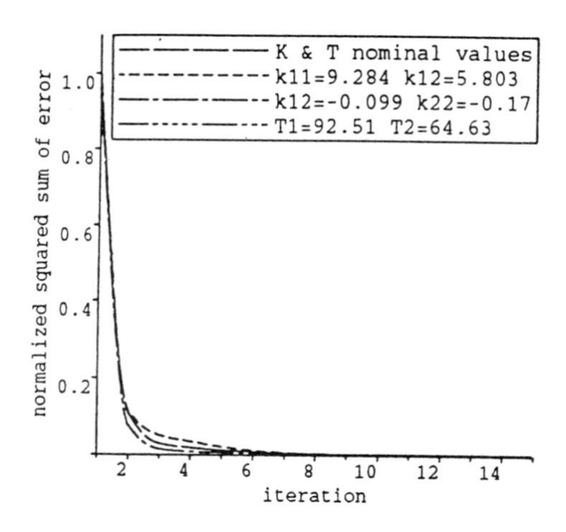


**Figure 10.7** Adaptive ability to process parameters

It can be seen that after a few iterations, the normalized SSE (NSSE) tend to stabilize at small values, indicating that the controller is able, in a uniform manner, to follow different performance requirements and to adapt to the different process parameters.

Finally, we examined the learning convergence property of the system with respect to various controller parameters $p_L$, $\beta$, $\delta$, and $\phi_0$. As shown in Figure 10.8, the learning gain matrix $P_L$ affects the learning speed more than the learning rate $\beta$.

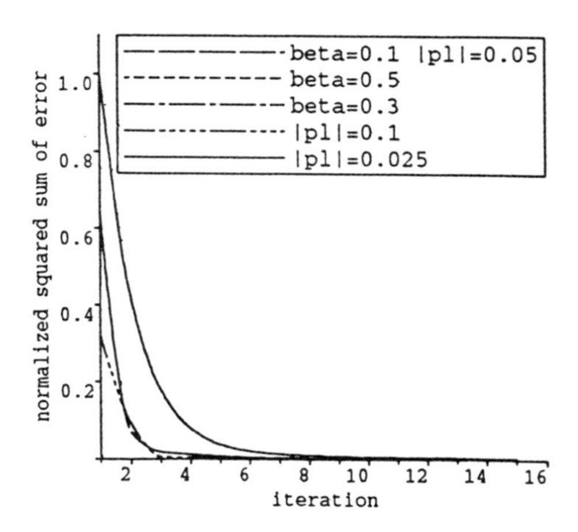


**Figure 10.8** Convergence to the learning rates

As would be expected, while the learning convergence is not very sensitive to the choice of the width $\delta$ and threshold $\phi_0$, the number of created hidden units (rules), however, is directly related to these parameters. Figure 10.9 shows the created hidden units versus iteration with different $\delta$ and $\phi_0$. We see that starting from a large number of units, the number of hidden units approaches gradually stable and small values. A general conclusion is that a bigger $\delta$ produces fewer units, and a bigger $\phi_0$ gives more units, conclusions which are identical to those from conceptual considerations.

## 10.6 Summary

We have described an approach of how a localized network can be fuzzified into a class of more general networks FLN. In particular, two well-known networks, the CMAC and the RBF, have been treated in detail. Furthermore, we have extended this to self-constructing systematically a FLN-based multivariable controller via learning.

Compared with the traditional CMAC, the FLN has the following features: (a)

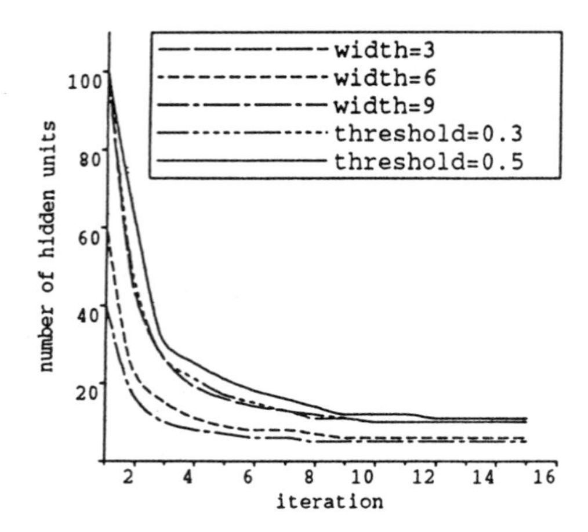


**Figure 10.9** Created hidden units versus iteration

the graded neighbourhood function overcomes the problem of discontinuous responses over neighbourhood boundaries; (b) the matching degree algorithm provides a simple similarity-measure-based content-addressable scheme capable of dealing with arbitrary-dimensional continuous input space without involving complicated quantizing, addressing, and hashing procedures as used in the original CMAC whose complexity increases with the input space dimension so dramatically that the system may become intractable.

On the other hand, the following conclusions can be drawn in connection the FLN to the traditional RBF networks: (a) the matching degree interpretation of the basis function has suggested that the basis function can not only be made factorable but can also be readily synthesized from $n$ independent localized functions by fuzzy logic operators, this having some significant implications, particularly for hardware implementations; (b) the correspondence between basis functions in RBF and membership functions in fuzzy systems provides an alternative explanation of why the choice of basis functions is not very sensitive with respect to system performance, a conclusion which has also been verified by the simulation; (c) the analogy between SFCA and RBF suggests that fuzzy systems like RBF possess a capacity for universal approximation.

Finally, the proposed control system structure and corresponding learning algorithms offer a simple and systematic approach to self-organizing the control knowledge with arbitrary but physically achievable performance requirements, and which can be implemented in real-time. Simulation results have demonstrated the feasibility of the method and provided some insight into the properties of the system.

CHAPTER 11

# Conclusions and further work

---

*We have dealt extensively with three key issues of knowledge representation, reasoning, and acquisition, relevant to the implementation of intelligent control systems under fuzzy algorithm-based and fuzzy-neural-based frameworks. Considerable effort has been dedicated to developing systems capable of performing self-organizing and self-learning functions in a real-time manner under multivariable system requirements with emphasis on the utilization of minimum prior knowledge about the process being controlled. To conclude this book, in what follows we will first summarize the main contributions of this work, then give some comparative comments on the developed learning systems, and finally suggest some points relevant to future work along the current research lines.*

---

## 11.1 Main contributions

The main contributions of this work are summarized as follows:

- We have shown how fuzzy systems can be studied systematically under the headings of *representation, reasoning,* and *acquisition*, terms used in general AI systems at the symbolic or linguistic level, and further how these issues can be recast techniquely into more practical issues, being appropriately termed as *structure, computation*, and *learning* for implementing algorithm-based and network-based fuzzy control systems at the numerical level.
- Within the scope of fuzzy algorithm-based schemes, we have developed a unified approximate reasoning model suitable for embedding various definitions of linguistic connectives, for handling possibilistic and probabilistic uncertainties,

and for dealing with linguistic and numerical environments. The model possesses desirable and useful properties of flexible and general applicability, meaningful interpretation of its operations, and retention of a parallel layer-like structure which can be readily linked with a kind of neural network. For the purpose of multivariable control applications, a viable decentralized fuzzy controller structure incorporating the established reasoning models has been proposed and demonstrated, giving a partial solution to the problem posed by interaction within variables.

- In terms of fuzzy-neural aspects, we have suggested two distinct approaches to merging fuzzy systems with neural systems and given detailed implementation methods and associated computational algorithms. It has revealed that fuzzy-logic algorithm-based reasoning systems working in numerical environments can be effectively and efficiently implemented by neural networks on the basis of either an *equivalence functional principle* or an *isomorphic structure rationale*, leading to either a distributed network structure such as a Back-propagation Neural Network or a localized network structure such as a Counterpropagation Neural Network. The above approaches and principles not only provide useful insight into a deeper understanding of the relationships existing naturally between fuzzy systems and neural networks, but also offer immediate solutions and further applicable and general guidelines to how fuzzy-neural systems can be practically perceived and developed.
- Again with respect to the fuzzy-neural structure, we have developed a class of fuzzified localization networks (FLN). Two well-known networks, CMAC and RBF, have been taken as examples of showing how this kind of three-layered networks can be generalized to incorporating fuzzy logic operators into the networks. The benefits due to the integration of fuzzy systems with neural networks have been highlighted.
- In an attempt to meet the crucial challenge of constructing automatically fuzzy controllers directly from the controlled environments instead of relying on domain experts, we have developed the five learning approaches reported in Chapters 5, 7, 8, 9 and 10. They are capable of meeting the above challenge, to some degree, in terms of self-learning and self-organizing. In particular, we have (1) provided some solid mathematical foundations for the learning mechanism (Chapter 4) which is utilized as an essential basis for developing self-learning schemes presented in Chapters 5, 7, and 9 respectively; (2) presented general principles, system structures, relevant algorithms, and detailed implementations with respect to each approach; (3) demonstrated the feasibility of the proposed strategies via extensive simulation studies of the problem of multivariable blood pressure control and explored their learning and adaptive properties with regard to a wide range of situations, thereby offering an understanding of how these vital properties are related to the parameters of the controller and the controlled process.

- This work has placed particular emphasis on the generality, simplicity, and systematicity in developing various design approaches with minimum knowledge requirements about the process being controlled. It has turned out that the resultant approaches dictated by these demands can be employed potentially to handle some relatively complicated multivariable control problems, in particular those problems where neither control experts nor mathematical models of the controlled process are available.

## 11.2 Comments on learning paradigms

It is worth while at this point to review the five learning paradigms (LP) developed in this book. For convenient reference hereafter, they will be quoted as $LP_{fla}$, $LP_{bnn}$, $LP_{hnn}$, $LP_{cpn}$, and $LP_{fln}$ corresponding to the learning paradigms presented in Chapters 5 (fuzzy logic algorithm-based), 7 (BNN network-based), 8 (hybrid network-based), 9 (CPN network-based), and 10 (FLN network-based) respectively. Some comments are as follows:

- All of them aim to construct rule-bases directly from the controlled process in accordance with arbitrary but physically achievable control performance requirements specified by reference models. In principle, all of them can cope with, in one way or another, situations of a process with an arbitrary number of variables and an associated controller with an arbitrary number of input variables.
- While $LP_{fla}$ and $LP_{bnn}$ involve three or four successive stages to complete the task, $LP_{hnn}$, $LP_{cpn}$ and $LP_{fln}$ do not need any intermittent procedures although they can indeed work with or without learning ability. In this sense, $LP_{hnn}$, $LP_{cpn}$ and $LP_{fln}$ are less complicated and more systematic than $LP_{fla}$ and $LP_{bnn}$.
- Regarding the algorithm-based or network-based fuzzy controllers as learning agents, the methodology adopted in $LP_{fla}$ and $LP_{bnn}$ can be considered as *passive off-line learning from examples*. In contrast, the strategy involved in $LP_{hnn}$, $LP_{cpn}$ and $LP_{fln}$ can be thought of as *active on-line learning from environments*. It is evident that the latter approach is more attractive, more efficient, and more important than the former.
- The rule-bases in $LP_{hnn}$, $LP_{cpn}$ and $LP_{fln}$ are built in a self-organizing manner, but $LP_{fla}$ and $LP_{bnn}$ do not share this feature. This can be seen by the fact that the size of the rule-bases and the IF part of each rule in $LP_{hnn}$, $LP_{cpn}$ and $LP_{fln}$ are determined automatically along with the process of learning whereas the determination of the above in $LP_{fla}$ and $LP_{bnn}$ relies on a rule-extracting procedure.
- It is useful to make some more specific comparisons between $LP_{hnn}$ and $LP_{cpn}$. While $LP_{hnn}$ uses a *BNN* network in which the underlying control rules are represented implicitly and distributedly, $LP_{cpn}$ built with localized CPN network exhibits the characteristic of explicit representation of a rule-base. Although the

network in $LP_{hnn}$ is a hybrid one incorporating a competitive network functioning partially as an explanation facility, this does unfortunately make the whole system more complex. In contrast, $LP_{cpn}$ has a much simpler topological structure. Another significant advantage of $LP_{cpn}$ over $LP_{hnn}$ is that the former provides a fast learning speed due to its localization structure and efficient learning algorithm which is linear in the parameters. Nevertheless, from the viewpoint of hardware realization, $LP_{hnn}$ possesses the merits of easier implementation and good fault tolerance. Moreover, the hybrid network structure in $LP_{hnn}$ provides a useful mechanism to facilitate efficiency of the BNN network which is widely used in many practical areas.

- Although $LP_{cpn}$ and $LP_{fln}$ share the same learning principle and the same systems structure, the detailed algorithm for constructing the network is different. While the minimum distance is used in $LP_{cpn}$ as the winner-selection criterion, $LP_{fln}$ employs the matching degree to determine the winner, leading to the utilization of more general and more flexible structures. Instead of using the winner-take-all exponential learning scheme for learning the output weight vector $\pi$ in $LP_{cpn}$, normally more than one $\pi$ vectors are updated in $LP_{fln}$ at each learning instant. Thus, $LP_{cpn}$ exhibits a more localized property and offers a faster learning speed.
- It seems clear that the above comments lead to a conclusion in favour of $LP_{hnn}$, $LP_{cpn}$ and $LP_{fln}$ paradigms. Depending on the circumstances, however, $LP_{fla}$ and $LP_{bnn}$ may find their own usefulness in those situations where the separate stages are essential and necessary. For example, it is possible to derive a rule-base embedding more global control knowledge for a specific controlled process by extracting the rule-bases from a relatively large set of data obtained from performing the learning stage many times with respect to a wide range of relevant operating conditions.

## 11.3 Future work

It is evident that fuzzy logic and neural networks are playing more and more significant roles in developing intelligent systems. The work reported in this book has shown some promising and encouraging results particularly in applying integrated fuzzy-neural networks to control problems. Despite significant achievements in this work, we feel that there are still many topics either untouched or not explored deeply enough, and restrictions imposed on the proposed approaches. For example, there is a lack of thorough analysis results pertinent to stability and/or learning convergence of the closed-loop systems. The assumption of a monotonic process is a typical restriction. Furthermore, it is not clear to what extent the proposed approaches possess the power to deal with more challenging and demanding control problems where the controlled process, besides its multivariable complexity, may be open-loop unstable, nonminimum phase, and strongly nonlinear. The above points can be regarded as part of future work deserving further investigation.

A particular point we have already considered is the control of nonlinear systems. We have applied the modified CPN network-based controller described in Chapter 9 to a problem of nonlinear multivariable anaesthesia control. It involves a simultaneous control of relaxation and unconsciousness during anaesthesia by using two drugs: atracurium and isoflurane. The system is characterized by interaction between variables and strong nonlinearity existing in the first channel. The overall simulation model has been experimentally justified and is given in Linkens *et al.* (1992). Preliminary results are encouraging. Figure 11.1 shows one of the results, indicating an acceptable performance.

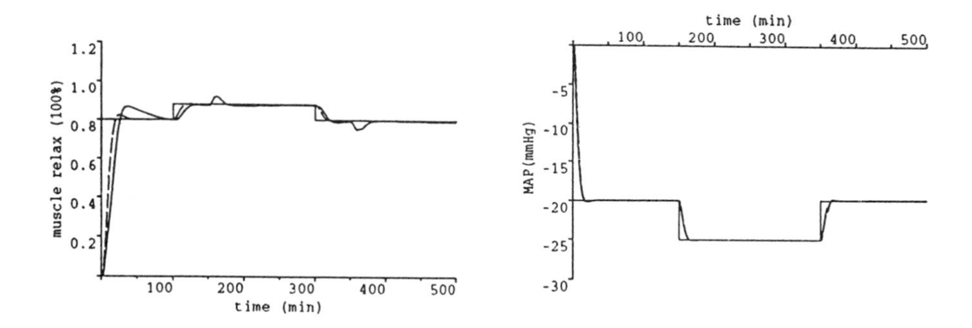


**Figure 11.1** Nonlinear anaesthesia control

It is equally true that there indeed exist many interesting and worthwhile issues along with the research lines adopted in this work, in particular that of combining fuzzy systems with neural networks in terms of developing more powerful representational structures and more efficient learning algorithms so as to meet the increasing practical demands relevant not only to control problems at a numerical level but also to general decision-making problems at a linguistic level.

APPENDIX I

# Moller's model of the cardiovascular system

Over the years, a variety of mathematical models of the cardiovascular system ($CVS$) have been developed (Mansour 1988), some of which are pulsatile and are suitable for investigating phenomena that can change in a fraction of a heart beat. Models for studying long-term effects are usually nonpulsatile. In this case blood circulation is described in terms of non-pulsatile pressures, flows and volumes. The model used in this study is a nonpulsatile model developed by Moller *et al.* (1983). A two-pump circulatory model was postulated and the basic relationship governing the physiologically closed $CVS$ was derived. The two parts of the heart are represented by flow sources $QL$ and $QR$ for left and right ventricles respectively. The systemic circulation is represented by the arterial systemic compliance $CAS$, the systemic peripheral resistance $RA$, and the venous systemic distensible capacity $CVS$. Similarly, the pulmonary circulation consists of the arteriopulmonary compliance $CAP$, the pulmonary resistance $RP$, and the venopulmonary distensible capacity $CVP$. The cardiovascular system dynamics governing the relationship between the blood pressures and flow sources can be described by the following differential vector equation

$$\mathrm{d}\mathbf{X}/\mathrm{d}t = A\,\mathbf{X} + B\,\mathbf{V} \tag{A.1}$$

where $\mathbf{X} = [PAS, PVS, PAP, PVP]^T$ with $PAS$, $PVS$, $PAP$ and $PVP$ being the systemic arterial pressure, systemic venous pressure, pulmonary arterial pressure and pulmonary venous pressure respectively; $\mathbf{V} = [QL, QR]^T$; and

$$A = \begin{bmatrix} (-CAS\cdot RA)^{-1} & (CAS\cdot RA)^{-1} & 0 & 0 \\ (CVS\cdot RA)^{-1} & (-CVS\cdot RA)^{-1} & 0 & 0 \\ 0 & 0 & (-CAP\cdot RP)^{-1} & (CAP\cdot RP)^{-1} \\ 0 & 0 & (CVP\cdot RP)^{-1} & (-CVP\cdot RP)^{-1} \end{bmatrix}$$

$$B = \begin{bmatrix} 1/CAS & 0 \\ 0 & -1/CVS \\ 0 & 1/CAP \\ -1/CVP & 0 \end{bmatrix}$$

Equation (A.1) is a nonlinear vector equation because the resistances and the compliances are nonlinear functions of the pressure. Furthermore, the inputs $QL$ and $QR$ are also pressure-dependent, that is,

$$QL = SV_L \cdot HR \qquad\qquad QR = SV_R \cdot HR \tag{A.2}$$

where $HR$ stands for the heart rate, and $SV_L$ and $SV_R$ are the stroke volumes of left and right ventricles respectively. $SV$ can be related to arterial and venous pressures by a complicated nonlinear algebraic function (Moller *et al.* 1983).

If the compliances and the resistances are treated as pressure-independent, equation (A.1) represents a linear state-space model with $QL$ and $QR$ as independent system inputs. By selecting the heart rate $HR$ and resistance $RA$ as the system inputs, Moller *et al.* derived a linear model near the stationary state $PAS_0 = 117.5$ mmHg, $PVS_0 = 7.15$ mmHg, $PAP_0 = 17.18$ mmHg and $PVP_0 = 10.87$ mmHg, which is given by

$$\mathrm{d}\Delta\mathbf{X}/\mathrm{d}t = A_l \Delta\mathbf{X} + B_l \Delta\mathbf{U} \tag{A.3}$$

where $\Delta\,\mathbf{X} = [\Delta PAS, \Delta PVS, \Delta PAP, \Delta PVP\,]^T$, $\Delta\,\mathbf{U} = [\Delta HR, \Delta RA\,]^T$ and

$$A_l = \begin{bmatrix} -3.4370 & 1.8475 & 0.0 & 18.7584 \\ 0.01834 & -0.3015 & 0.06855 & 0.0 \\ 0.0 & 7.3514 & -10.1131 & 8.3333 \\ 0.2049 & 0.0 & 4.1667 & -6.5846 \end{bmatrix}$$

$$B_l = \begin{bmatrix} 125.8 & 194.3 \\ -0.5048 & -1.929 \\ 13.1058 & 0.0 \\ -16.2125 & 0.0 \end{bmatrix}$$

Equation (A.3) indicates that the processes can be controlled with a fast response time by manipulating heart rate and systemic resistance. It should be noted that the activation of $HR$ is currently feasible through direct electrical stimulation of the heart, but is not yet available directly for $RA$.

APPENDIX II

# Model of drug dynamics

Simultaneous regulation of blood pressure and cardiac output ($CO$) is needed in some clinical situations, for instance congestive heart failure. It is desirable to maintain or increase $CO$ and, at the same time, to decrease the blood pressure. This goal can be achieved by simultaneous infusions of a positive inotropic agent, which increases the heart's contractility and cardiac output, and with a vasodilator which dilates the vasoculature and lowers the arterial pressure. Two frequently used drugs in clinical practice are the inotropic drug dopamine (DOP) and the vasoactive drug sodium nitroprusside (SNP). It is worth noting that the inputs are interactive with respect to the controlled variables $CO$ and mean arterial pressure ($MAP$). The inotropic agent increases $CO$ and thus $MAP$, whereas the vasoactive agent decreases $MAP$ and increases $CO$.

An accurate dynamical model associating $CO$ and $MAP$ with DOP and SNP is not available to date. However, Serna *et al.* (1983) derived a first-order model for which different time constants and time delays in each loop were obtained. The steady-state gains in the model were obtained from Miller's study (1977). The dynamics in the $s$-domain are given by

$$\begin{bmatrix} \Delta CO_d \\ \Delta RA_d \end{bmatrix} = \begin{bmatrix} K_{11}e^{-\tau_1 s}/sT_1+1 & K_{12}e^{-\tau_2 s}/sT_1+1 \\ K_{21}e^{-\tau_2 s}/sT_2+1 & K_{22}e^{-\tau_2 s}/sT_2+1 \end{bmatrix} \begin{bmatrix} I_1 \\ I_2 \end{bmatrix} \tag{A.4}$$

where $\Delta CO_d$ (ml/s) is the change in $CO$ due to $I_1$ and $I_2$; $\Delta RA_d$ (mmHg.s/ml) is the change in $RA$ due to $I_1$ and $I_2$; $I_1$ (µg/kg/min) is the infusion rate of DOP; $I_2$ (ml/h) is the infusion rate of SNP; $K_{11}$, $K_{12}$, $K_{21}$ and $K_{22}$ are steady-state gains with typical values of 8.44, 5.275, –0.09 and –0.15 respectively; $\tau_1$ and $\tau_2$ represent two time delays with typical values of $\tau_1 = 60$ s and $\tau_2 = 30$ s; $T_1$ and $T_2$ are time constants typified by the values of 84.1 s and 58.75 s respectively. The model parameters presented above will be varied during the simulations in order to evaluate the robustness of the proposed controller.

Because the accessible measurable variables are $MAP$ and $CO$, a model which relates the $\Delta CO_d$ and $\Delta RA_d$ due to drug infusions is needed. Moller's cardiovascular model can be used for this purpose. Note that the cardiovascular dynamics are much faster than the drug dynamics. Consequently, it is reasonable to neglect the cardiovascular dynamics, and only retain the steady state gains in the *CVS* model. With this consideration, Mansour and Linkens (1990) derived a simulation model from Moller's *CVS* model and the drug dynamics, which is given by

$$\begin{bmatrix} \Delta CO \\ \Delta MAP \end{bmatrix} = \begin{bmatrix} 1.0 & -24.76 \\ 0.6636 & 76.38 \end{bmatrix} \begin{bmatrix} K_{11}e^{-\tau_1 s}/sT_1{+}1 & K_{12}e^{-\tau_2 s}/sT_1{+}1 \\ K_{21}e^{-\tau_2 s}/sT_2{+}1 & K_{22}e^{-\tau_2 s}/sT_2{+}1 \end{bmatrix} \begin{bmatrix} I_1 \\ I_2 \end{bmatrix} \tag{A.5}$$

$$= \begin{bmatrix} H_{11} & H_{12} \\ H_{21} & H_{22} \end{bmatrix} \begin{bmatrix} I_1 \\ I_2 \end{bmatrix} \tag{A.6}$$

where

$$\begin{aligned} H_{11} &= 1.0 \cdot K_{11}e^{-\tau_1 s}/T_1 s+1 + -24.76 \cdot K_{21}e^{-\tau_2 s}/T_2 s+1 \\ H_{12} &= 1.0 \cdot K_{12}e^{-\tau_2 s}/T_1 s+1 + -24.76 \cdot K_{22}e^{-\tau_2 s}/T_2 s+1 \\ H_{21} &= .6636 \cdot K_{11}e^{-\tau_1 s}/T_1 s+1 + 76.38 \cdot K_{21}e^{-\tau_2 s}/T_2 s+1 \\ H_{22} &= 0.6636 \cdot K_{12}e^{-\tau_2 s}/T_1 s+1 + 76.38 \cdot K_{22}e^{-\tau_2 s}/T_2 s+1 \end{aligned} \tag{A.7}$$

It is evident that the model is characterized by strong interactions between variables and large time delays in control.

# Bibliography


An, P.C.E., Miller III, W.T., and Parks, P.C. (1991), Design improvements in associative memories for CMAC, *Proc. Int. Conf. on Artificial Neural Networks*, Helsinki: North-Holland, pp 1207-1210.

Albus, J.S. (1975a), A new approach to manipulator control: The Cerebellar Model Articulation Controller (CMAC), *J. Dynamic Syst. Meas. Contr.*, **97,** pp 220-227.

Albus, J.S. (1975b), Data storage in the Cerebellar Model Articulation Controller (CMAC), *J. Dynamic Syst. Meas. Contr.*, **97,** pp 228-233.

Anderson, C.W. (1989), Learning to control an inverted pendulum using neural networks, *IEEE Control Magazine,* pp 31-37, April.

Arimoto, S. (1986), Mathematical theory of learning with application to robot control, in, *Adaptive and Learning Systems* (K. S. Narendra ed.), New York: Plenum Press, 379-388.

Arimoto, S. (1990), Learning control theory for robotic motion, *Int. J. Adaptive Control and Signal Processing,* **4,** pp 543-564.

Arimoto, S., Kawamura S. and Miyazaki, F. (1984), Bettering operation of robots by learning, *J. Robotic Systems,* **1,** pp 123-140.

Arimoto, S., Kawamura S., Miyazaki, F. and Tamaki, S. (1985), Learning control theory for dynamical systems, *Proc. 24th IEEE Conf. on Decision and Control,* pp 1375-1380.

Astrom, K.J., Anton, J.J. and Arzen, K.E. (1986), Expert control, *Automatica*, **22**, pp 277-286.

Astrom, K.J. and Wittenmark, B. (1989), *Adaptive Control*, Reading, MA: Addison-Wesley.

Barto, A.G. (1990), Connectionist learning for control: an overview, in *Neural Networks for Control* (W.T. Miller *et al.* eds), Cambridge, MA: MIT Press, pp 5-58.

Barto, A.G., Sutton, R.S. and Anderson, C. W. (1983), Neurolike adaptive elements that can solve difficult control problems, *IEEE Trans. Systems, Man, and Cybernetics*, **13**, pp 834-846.

Baum, E.B. and Haussler, D. (1989), What size net gives valid generalization, *Neural Computation,* **1,** pp 151-160.

Bechtel, W. and Abrahamsen, A. (1991), *Connectionism and the Mind*, Oxford: Basil Blackwell.

Berenji, H.R. (1992), A reinforcement learning-based architecture for fuzzy logic control, *Int. J. Approximate Reasoning*, **6,** pp 267-292.

Bezdek, J.C. ed. (1992), Special issue on fuzzy systems, *IEEE Trans. Neural Networks*, **3**, September, pp 643-769.

Bezdek, J.C. and Pal, S.K. eds (1992), *Fuzzy Models for Pattern Recognition*, New York: IEEE Press.

Bien, Z. and Hah, K.M. (1989), Higher-order iterative learning control algorithm, *IEE Proc. Part D,* **136,** pp 105-112.

Broomhead, D.S. and Lowe, D. (1988), Multivariable functional interpolation and adaptive networks, *Complex Syst.*, **2,** pp 321-355.

Brown, M. and Harris, C.J. (1991), Fuzzy logic, neural networks and B-spline for intelligent control, *IMA J. Math. Control and Information Theory*, **8**, pp 239-265.

Brown, M. and Harris, C.J. (1992), Least mean square learning in associative memory networks, *IEEE Int Symp. on Intelligent Control,* pp 531-536.

Brown, M., Harris, C.J. and Parks, P.C. (1993), The interpolation capabilities of the binary CMAC, *Neural Networks*, **6,** pp 429-440.

Buckley, J.J., Hayashi, Y. and Czogala, E. (1993), On the equivalence of neural nets and fuzzy expert systems, *Fuzzy Sets and Systems*, **53,** pp 129-134.

Caudill, M. (1991), Expert network, *Byte,* **16,** October, pp 108-112.

Carpenter, G.A., Grossberg S. and Rosen, D.B. (1991), Fuzzy ART: Fast stable learning and categorization of analog patterns by an adaptive resonance system, *Neural Networks*, **4,** pp 759-771.

Chen, F.C. (1990), Back-propagation neural networks for nonlinear self-tuning adaptive control, *IEEE Control Systems Magazine*, pp 44-48.

Chen, S. and Billings, S.A. (1992), Neural networks for nonlinear dynamic system modeling and identification, *Int. J. Control*, **56,** pp 319-346.

Chen, Y.Y. (1989), Rule extraction for fuzzy control systems, *Proc. 1989 IEEE Int. Conf. on Syst. Man Cybern.,* **2,** pp 526-527.

Clark, D.M. and Ravishankar, K. (1990), A convergence theorem for Grossberg learning, *Neural Networks* **3,** pp 87-92.

Cotter, N.E. and Guillerm, T.J. (1992), The CMAC and a theorem of Kolmogorov, *Neural Networks*, **5,** pp 221-228.

Daley, S. and Gill, K.F. (1986), A design study of a self-organizing fuzzy logic controller, *Proc. Instn. Mech. Engrs.,* **200,** pp 59-69.

den Brok, M.W.N.M. and Blom, J.A. (1987), A rule-based adaptive blood pressure controller. *Proc. of the Second European Workshop on Fault Diagnostics, Reliability and Related Knowledge-based Approaches,* pp 67-74.

Desoer, C.A. and Vidyasagar, M. (1975), *Feedback Systems: Input-output Properties*, New York: Academic Press.

Duda, R.O. and Hart, P.E. (1973), *Classification and Scene Analysis*, New York: Wiley.

Efstathiou, J. (1988), Expert systems, fuzzy logic and rule-based control explained at last. *Trans Inst MC,* **10,** pp 198-206.

Efstathiou, J. (1989), *Expert Systems in Process Control*, Harlow: Longman, in association with the Institute of Measurement and Control.

Ellison, D. (1991), On the convergence of the multidimensional Albus perceptron, *Int. J. Robotics Research*, pp 338-357.

Eppler, W. (1990), Implementation of fuzzy production systems with neural networks, in *Parallel Processing in Neural Systems and Computers* (R. Eckmiller ed.), New York: Elsevier Science, pp 249-252.

Eshragh, F. and Mamdani, E.H., (1981), A general approach to linguistic approximation, in *Fuzzy Reasoning and its Application* (E. H. Mamdani and B. R. Gaines eds), New York: Academic Press.

Fu, L.M. (1989), Integration of neural heuristics into knowledge-based inference, *Connection Science,* **1,** pp 325-340.

Funahashi, K. (1989), On the approximate realization of continuous mappings by neural networks, *Neural Networks*, **2,** pp 183-192.

Gallant, S. (1988), Connectionist expert systems, *Communication of the ACM,* **31,** pp 152-169.

Geng, Z., Carroll, R. and Xie, J.H. (1990), Two-dimensional model and algorithm analysis for a class of iterative learning control systems, *Int. J. Control,* **52,** pp 833-862.

Grosdidier, P. and Morari, M. (1985), Closed-loop properties from steady-state gain information. *Industrial and Engineering Chemistry Fundamentals,* **24,** pp 221-235.

Grossberg, S. (1987), Competitive learning: From interactive activation to adaptive resonance, *Cognitive Science*, **11**, pp 23-63.

Gupta, M.M. *et al.* eds (1985), *Approximate Reasoning in Expert Systems*, Amsterdam: North-Holland.

Gupta, M.M., Kiszka, J.B. and Trojan, G.M. (1986), Multivariable structure of fuzzy control systems. *IEEE Trans. Syst. Man Cybern.,* **16,** pp 638-656.

Gutknecht, M. and Pfeifer, R. (1990), An approach to integrating expert systems with connectionist networks, *AICOM,* **3,** pp 116-127.

Hara, S., Yamamoto, Y., Omata, T. and Nakano, M. (1988), Repetitive control system: a new type servo system for periodic exogenous signals, *IEEE Trans. Automatic Control,* **33,** pp 659-667.

Harris, C.J. (1992a), Comparative aspects of neural networks and fuzzy logic for real-time control, in *Neural Networks for Control and Systems* (K. Warwick *et al.* eds), Stevenage: Peter Peregrinus Ltd.

Harris, C.J. (1992b), Guest editorial for special issue on intelligent control, *Int. J. Control,* **56,** pp 259-261.

Harris, C.J. and Moore, C.G. (1989), Intelligent identification and control for autonomous guided vehicles using adaptive fuzzy based algorithms, *Int. J. Engr. Application of AI,* **2,** pp 267-285.

Harris, C.J., Moore, C.G. and Brown, M. (1992), *Intelligent Control: Some Aspects of Fuzzy Logic and Neural Networks,* London and Singapore: World Scientific Press.

Hecht-Nielsen, R. (1987), Counterpropagation network, *Applied Optics*, **26**, pp 4979-4984.

Hecht-Nielsen, R. (1988), Applications of counterpropagation network, *Neural Networks*, **1**, pp 131-139.

Hecht-Nielsen, R. (1990), *Neurocomputing*, Reading, MA: Addison-Wesley.

Holmblad, L.P. and Ostergaard, J.J. (1982), Control of a cement kiln by fuzzy logic, in *Fuzzy Information and Decision Processes* (M. M. Gupta *et al.* eds), Amsterdam: North-Holland.

Horikawa, S., Furuhashi, T. and Uchikawa, Y. (1992), On fuzzy modelling using fuzzy neural networks with back-propagation algorithm, *IEEE Trans. Neural Networks*, **3,** pp 801-806.

Hornic, K. (1991), Approximate capabilities of multilayer feedforward networks, *Neural Networks,* **4,** pp 251-257.

Hornic, K., Stinchcombe, M. and White, H. (1989), Multilayer feedforward networks are universal approximators, *Neural Networks,* **2,** pp 359-366.

Hwang, D.H., Bien, Z. and Oh, S.R. (1991), Iterative learning control method for discrete-time dynamic, *IEE Proc. Part D,* **138,** pp 139-144.

Ichikawa, Y. and Sawa, T. (1992), Neural network applications for direct feedback controller, *IEEE Trans. Neural Networks,* **3,** pp 224-231.

Isaka, S. *et al.* (1988), On the design and performance evaluation of adaptive fuzzy controller. *Proc. IEEE the 27th Conference on Decision and Control,* 1068-1069.

Isaka, S. *et al.* (1989), An adaptive fuzzy controller for blood pressure regulation. *Proc. IEEE Engineering in Medicine and Biology Society 11th Annual Int. Conf.,* 1763-1764.

Jensen, N., Fisher, D.G. and Shah, S.L. (1986), Interaction analysis in multivariable control systems. *AIchE Journal,* **32,** pp 959-970.

Jordan, M. (1989), Generic constraints on underspecified target trajectories, *Proc. of Int. Joint. Conf. on Neural Networks,* pp 217-225.

Jordan, M. and Rumelhart, D. (1992) Forward models: supervised learning with a distal teacher, *Cognitive Science*, **16,** pp 307-354.

Kawato, M. (1990), Computational schemes and neural network models for formation and control of multijoint trajectory, in *Neural Networks for Control* (W. T. Miller *et al.* eds), Cambridge, MA: MIT Press, pp 197-228.

Keller, J., Yager, R. and Tahani, H. (1992), Neural network implementation of fuzzy logic, *Fuzzy Sets and Systems*, **45**, pp1-12.

Kohonen, T. (1988a), *Self-organizing and Associative Memory*, Berlin: Springer-Verlag.

Kohonen, T. (1988b), The neural phonetic typewriter, *IEEE Computer Magazine,* March, pp 11-22.

Kohonen, T. (1990), Statistical pattern recognition revised, in *Advanced Neural Computers* (R. Eckmiller ed.), Amsterdam: North-Holland, pp 137-144.

Kong, S.G. and Kosko, B. (1992), Adaptive fuzzy systems for backing up a truck-and-trailer, *IEEE Trans. Neural Networks,* **3,** pp 211-233.

Kosko, B. (1991), Stochastic competitive learning, *IEEE Trans. Neural Networks*, **2**, pp 522-529.

Kosko, B. (1992a), Fuzzy system as universal approximators, *Proc. of IEEE Int. Conf. on Fuzzy Systems*, pp 1153-1162.

Kosko, B. (1992b), *Neural Networks and Fuzzy Systems*, Englewood Cliffs, NJ: Prentice Hall.

Kottai, R. and Bahill, A.T. (1989), Expert systems made with neural networks, *Int. J. Neural Networks,* **1,** pp 211-226.

Lacher, R.C. (1992), Back-propagation learning in expert networks, *IEEE Trans. Neural Networks,* **3,** pp 62-72.

Lane, S.H., Handelman, D.A. and Gelfand, J.J. (1992), Theory and development of high-order CMAC neural networks, *IEEE Control Systems Magazine*, **12,** pp 23-30.

Lee, C.C. (1990), Fuzzy logic in control systems: fuzzy logic controller part one and part two, *IEEE Trans. Syst. Man Cybern.,* **20,** pp 404-435.

Lee, C.C (1991), A self-learning rule-based controller employing approximate reasoning and neural net concepts, *Int. J. Intelligent Systems,* **6,** pp 71-93.

Lembessis, E. (1984), *Dynamic Learning Behaviour of a Rule-based Self-organizing Controller*, PhD thesis, QMC, University of London.

Lemmon, M.D. (1991), *Competitive Inhibited Neural Networks for Adaptive Parameter Estimation*, Boston, MA: Kluwer Academic Publishers.

Lim, M. H. and Takefuji, Y. (1990), Implementing fuzzy rule-based systems on silicon chips, *IEEE Expert,* pp 31-45, February.

Linkens, D.A. and Abbod, M.F. (1992), Self-organizing fuzzy logic control and the selection of its scaling factors, *Trans. Inst. MC*, **14**, pp 114-125.

Linkens, D.A., Greenhow, S.G., and Asbury, A.J. (1986), An expert system for the control of depth of anaesthesia, *Biomed. Meas. Infor. Contr.*, **1**, pp 223-228.

Linkens, D.A. and Mahfouf, M. (1988), Fuzzy logic knowledge-based control for muscle relaxant anaesthesia, *Proc. of IFAC Modelling and Control in Medicine*, pp 185-190.

Linkens, D.A., Mahfouf, M. and Asbury, A.J. (1991), Multivariable generalized predictive control for anaesthesia, *European Control Conference*, **2**, Grenoble, pp 1630-1635.

Linkens, D.A., Mahfouf, M. and Abbod, M.F. (1992), Self-adaptive and self-organizing control applied to nonlinear multivariable anaesthesia, *IEE Proc. D*, **139**, pp 131-139. pp 131-139.

Linkens, D.A. and Mansour, N.-E. (1989), Pole-placement self-tuning control of blood pressure in post-operative patients: a simulation study, *IEE Proc. D*, **136**, pp 1-11.

Linkens, D.A. and Nie, Junhong (1992a), A unified real time approximate reasoning approach for use in intelligent control. Part 1: theoretical development, *Int. J. Control,* **56**, pp 347-364.

Linkens, D.A. and Nie, Junhong (1992b), A unified real time approximate reasoning approach for use in intelligent control. Part 2: application to multivariable blood pressure control, *Int. J. Control,* **56**, pp 365-398.

Linkens, D.A. and Nie, Junhong (1992c), Rule extraction for BNN neural network-based fuzzy control systems by self-learning, in *Artificial Neural Networks 2* (I. Aleksander and J. Taylor eds), Amsterdam: Noth-Holland, pp 459-462.

Linkens, D.A. and Nie, Junhong (1993a), Constructing rule-bases for multivariable fuzzy control by self-learning. Part 1: system structure and learning algorithms, *Int. J. System Science*, **24**, pp 111-127.

Linkens, D.A. and Nie, Junhong (1993b), Constructing rule-bases for multivariable fuzzy control by self-learning. Part 2: Rule-base formation and blood pressure control application, *Int. J. System Science*, **24**, pp 129-167.

Linkens, D.A. and Nie, Junhong (1994), Back-propagation neural network-based fuzzy controller with self-learning teacher, *Int. J. Control*, **60**, pp 17-39.

Low, B.T., Lui, H.C, Tan, A.H. and Teh, H.H. (1991), Connectionist expert system with adaptive learning capability, *IEEE Trans. Knowledge and Data Eng.*, **3**, pp 200-207.

Maiers, J. and Sherif, Y.S. (1985), Application of fuzzy set theory, *IEEE Trans. Systems Man and Cyber.*, **15**, pp 175-189.

Mamdani, E.H. (1974), Application of fuzzy algorithms for control of simple dynamic plant, *Proc. IEE*, **121,** pp 1585-1588.

Mamdani, E.H. (1977), Applications of fuzzy set theory to control systems--a survey, in *Fuzzy Automata and Decision Process* (M. M. Gupta *et al.* eds), Amsterdam: North-Holland, pp 77-88.

Mamdani, E.H. (1993), Twenty years of fuzzy control: experiences gained and lessons learnt, *Proc. of IEEE Conf. on Fuzzy Systems*, pp 339-344.

Mamdani, E.H. and Assilian, S. (1975), An experiment in linguistic synthesis with a fuzzy logic controller, *Int. J. Man-Machine Studies,* **7,** pp 1-13.

Mamdani, E.H., Efstathiou, H.J. and Sugiyama, K. (1984), Developments in fuzzy logic control, *Proc. 23rd IEEE Conf. on Decision and Control*, pp 883-893.

Mamdani, E.H., Ostergaard, J.J. and Lembesis, E. (1984), Use of fuzzy logic for implementing rule-based control of industrial processes, *Tims Studies in Management Science*, **20**, pp 429-445.

Mansour, N.E. (1988), *Adaptive Control of Blood Pressure*, PhD thesis, Dept. of Automatic Control and Sys. Engineering, University of Sheffield.

Mansour, N.E. and Linkens, D.A. (1990), Self-tuning pole-placement multivariable control of blood pressure for post-operative patients: a model-based study. *IEE Proc. D*, **137,** pp 13-29.

Mcinnis, B.C. and Deng, L.Z. (1985), Automatic control of blood pressure with multiple drug inputs. *Ann. Biomed. Eng.*, **13,** pp 217-225.

Messner, W., Horowitz, R., Kao, W.W. and Oals, M. (1991), A new adaptive learning rule, *IEEE Trans. on Automatic Control,* **36,** pp 188-197.

Miller, R.R. *et al.* (1977), Combined dopamine and nitroprusside therapy in congestive heart failure, *Circulation,* **55,** p 881.

Miller, W.T. (1989), Real-time application of neural network for sensor-based control with vision, *IEEE Trans. Systems, Man, and Cybernetics*, **19**, pp 825-831.

Miller, W.T., Glanz, F.H. and Kraft, L.G. (1990), CMAC: an associative neural network alternative to back-propagation, *Proc. IEEE,* **78,** pp 1561-1567.

Miller, W.T, An, P.C.E., Glanz, F.H. and Carter, M.J. (1990), The design of CMAC neural networks for control, *Proc. Yale Workshop on Adaptive Systems.*

Minsky, M.L. (1991), Logical versus analogical or symbolic versus connectionist or neat versus scruffy, *AI Magazine,* Summer, pp 35-51.

Minsky, M.L. and Parpert, S.A. (1988), *Perceptrons*, 3rd edition, Cambridge, MA: MIT Press.

Mizumoto, M. and Zimmermann, H. (1982), Comparison of fuzzy reasoning methods, *Fuzzy Sets and Systems,* **8,** pp 253-283.

Mizumoto, M. (1987), Fuzzy control under various approximate reasoning methods, *Preprints of Second IFSA Congress,* pp 143-146.

Moller, D. *et al.* (1983), Modelling, simulation and parameter-estimation of the human cardiovascular system, in *Advances in control Systems and Signal Processing*, **4**, Braunschweig: Vieweg.

Moody, J. (1989), Fast learning in multi-resolution hierarchies, *Advances in Neural Information Processing Systems*, **1** (D. Touretzky ed.), Los Altos, CA: Morgan Kaufman.

Moody, J. and Darken, C. (1989), Fast-learning in networks of locally-tuned processing units, *Neural Computation*, **1**, *pp 281-294.*

Moore, C.G. and Harris, C.J. (1992), Indirect fuzzy control, *Int. J. Control*, **56**, pp 441-468

Narendra, K.S. (1990), Adaptive control using neural networks, in *Neural Networks for Control* (W. T. Miller *et al.* eds), Cambridge, MA: MIT Press, pp 115-142.

Narendra, K.S. and Parthasarthy, K. (1990), Identification and control of dynamical systems using neural networks, *IEEE Trans. Neural Networks*, **1**, pp 4-27.

Neat, G.W. *et al.* (1989), Expert adaptive control for drug delivery systems, *IEEE Control Systems Magazine*, June, pp 20-23.

Nerrand, O., Roussel-Ragot, P., Personnaz, L. and Dreyfus, G. (1993), Neural networks and nonlinear adaptive filtering: Unifying concepts and new algorithms, *Neural Computation*, **5**, pp 165-199.

Nguyen, D.H. and Widrow, B. (1990), Neural networks for self-learning control systems, *IEEE Control Systems Magazine*, pp 18-23.

Nie, Junhong (1987), *Expert fuzzy control systems*, MS thesis, Xidian University, Xi'an, China.

Nie, Junhong (1989), A class of new fuzzy control algorithms, *Proc. IEEE Int. Conf. on Control and Applications,* Israel.

Nie, Junhong and Linkens, D.A. (1992a), Fuzzy reasoning implemented by neural networks, *Proc. Int. Joint Conference on Neural Networks* (IJCNN'92), Baltimore, MD.

Nie, Junhong and Linkens, D.A. (1992b), Neural network-based approximate reasoning: principles and implementation, *Int. J. Control,* **56**, pp 399-413.

Nie, Junhong and Linkens, D.A. (1992c), Counterpropagation network-based fuzzy controllers: explicit representation and self-construction of rule-bases, in *Artificial Neural Networks 2* (I. Aleksander and J. Taylor eds), Amsterdam: Noth-Holland, pp 463-46.

Nie, Junhong and Linkens, D.A. (1993a), Automatic knowledge acquisition for multivariable fuzzy control using neural network approach, *Proc. of American Control Conference.*, pp 767-771.

Nie, Junhong and Linkens, D.A. (1993b), Learning control using fuzzified self-organizing radial basis function network, *IEEE Trans. Fuzzy Systems*, **1**, pp 280-287.

Nie, Junhong and Linkens, D.A. (1994a), A hybrid neural network-based self-organizing controller, *Int. J. Control,* **60**, pp 197-222.

Nie, Junhong and Linkens, D.A. (1994b), Fast self-learning multivariable fuzzy controllers built with modified CPN networks, *Int. J. Control,* **60**, pp 369-393.

Nie, Junhong and Linkens, D.A. (1994c), FCMAC: a fuzzified Cerebellar Model Articulation Controller with self-organizing capability, *Automatica,* **30**, pp 655-664.

Pao, Y.H. and Sobajic, D.J. (1991), Neural networks and knowledge engineering, *IEEE Trans. Knowledge and Data Engineering,* **3**, pp 185-192.

Park, J. and Sandberg, I.W. (1991), Universal approximation using radial-basis-function networks, *Neural Computation*, pp 246-257.

Park, J. and Sandberg, I.W. (1993), Approximation and radial-basis-function networks, *Neural Computation*, **5**, pp 305-316.

Parks, P.C. and Militzer, J. (1989), Convergence properties of associative memory storage for learning control systems, *Automatic Remote Control,*, **50**, pp 254-286.

Parks, P.C. and Militzer, J. (1991), Improved allocation of weights for associative memory storage in learning systems, *Proc. IFAC Symposium on Design Methods of Control Systems*, Zurich, Pergamon Press, pp 565-572.

Parks, P.C. and Militzer, J. (1992), A comparison of five algorithms for the training of CMAC memories for learning control systems, *Automatica*, **28,** pp 1027-1035.

Pedrycz, W. (1985a), Application of fuzzy relational equation for methods of reasoning in presence of fuzzy data, *Fuzzy Sets and Systems,* **16,** pp 163-175.

Pedrycz, W. (1985b), Design of fuzzy control algorithms with the aid of fuzzy models, in *Industrial Applications of Fuzzy Control* (M. Sugeno ed.), Amsterdam: North-Holland.

Poggio, T. and Girosi, F. (1990), Networks for approximation and learning, *Proc. IEEE,* **78,** pp 1481-1497.

Poter, B. and Mohamed, S. (1991), Iterative learning control of partially irregular multivariable plants with initial state shifting, *Int. J. Systems Scince.*

Procyk, T.J. and Mamdani, E.H. (1979), A linguistic self-organizing process controller, *Automatica,* **15,** pp 15-30.

Psaltis, D., Sideris, A. and Yamamura, A. (1988), A multilayered neural network controller, *IEEE Control Magazine,* April, pp 17-21.

Ray, K.S. *et al.* (1985), Structure of an intelligent fuzzy logic controller and its behaviour, in *Approximate Reasoning in Expert Systems* (M. M. Gupta *et al.* eds), Amsterdam: Elsevier Science, pp 553-561.

Rhee, F.V.D. *et al.* (1990), Knowledge based fuzzy control of systems, *IEEE Trans. on Automatic Control,* **35,** pp 148-155.

Rumelhart, D.E. and Zipser, D., 1985, Feature discovery by competitive learning, *Cognitive Science*, **9**, pp 75-112.

Rumelhart, D.E. Hinton, G.E. and Williams, R.J. (1986), Learning internal representation by error propagation, in *Parallel Distributed Processing* , **1** (D. E. Rumelhart and J. L. McClelland Eds), Cambridge, MA: MIT Press.

Sadegh, N., Horowitz, R., Kao, W.W. and Tomizuka, M. (1990), A unified approach to the design of adaptive and repetitive controllers for robotic manipulators, *J. of Dynamic Systs. Measur. and Contr.,* **112,** pp 618-629.

Sanner, R.M. and Akin, D.L. (1990), Neuraomorphic pitch attitude regulation of an underwater telerobot, *IEEE Control Systems Magazine*, April, pp 62-67.

Scharf, E.M. and Mandic, N.J. (1985), The application of a fuzzy controller to the control of a multi-degree-freedom robot arm, in *Industrial Applications of Fuzzy Control* (M. Sugeno ed.), Amsterdam: North-Holland.

Schwartz, D.G. and Klir, G.J. (1992), Fuzzy logic flowers in Japan, *IEEE Spectrum*, **29**, No.7, pp 32-35.

Serna, V. *et al.* (1983), Adaptive control of multiple drug infusion. *Proc. of JACC Conf.,* pp 22-26.

Shao, S. (1988), Fuzzy self-organizing controller and its application for dynamic processes, *Fuzzy Sets and Systems,* **26,** pp 151-164.

Simpson, P (1990), *Artificial Neural Systems,* Elmsford, NY: Pergamon Press.

Sugeno, M. (ed.) (1985), *Industrial Applications of Fuzzy Control*, Amsterdam: North-Holland.

Sugeno, M. and Kang, G.T. (1988), Structure identification of fuzzy model, *Fuzzy sets and systems,* **28,** pp 15-33.

Sugeno, M. and Nishida, M. (1985) Fuzzy control of model car, *Fuzzy Sets and Systems,* **16,** pp 103-113.

Sugeno, M. and Takagi, T. (1983), Multi-dimensional fuzzy reasoning, *Fuzzy Sets and Systems,* **9,** pp 313-325.

Sugie, T. and Ono, T. (1991), An iterative learning control law for dynamical systems, *Automatica,* **27,** pp 729-732.

Sugiyama, K. (1986), *Analysis and Synthesis of the Rule-based Self-organizing Controller*, PhD thesis, QMC, University of London.

Takagi, H. and Hayashi, A. (1991), NN-driven fuzzy reasoning, *Int. J. Approximate Reasoning,* **5,** pp 191-212.

Takagi, T. and Sugeno, M. (1985), Fuzzy identification of systems and its application to modeling and control, *IEEE Trans. Syst. Man Cybern.,* **15,** pp 116-132.

Tanscheit, R. (1988), *Study of a Rule-based Self-organizing Controller for Robotics Applications*, PhD thesis, QMC, University of London.

Togai, M. and Watanable, H. (1986), Expert system on chip: an engine for real-time approximate reasoning, *IEEE Expert,* Fall, pp 55-62.

Togai, M and Yamano, O. (1985), Analysis and design of an optimal learning control scheme for industrial robots, *Proc. of IEEE 24th Conference on Decision and Control,* pp 1399-1404.

Tolle, H. and Ersu, E. (1992), *Neural Control*, Berlin: Springer-Verlag.

Tolle, H., Parks, P.C., Ersu, E., Hormel, M. and Militzer, J. (1992), Learning control with interpolating memories: general ideas, design lay-out, theoretical approaches and applications, *Int. J. Control*, **56,** pp 291-317.

Tong, R.M. (1984), A retrospective view of fuzzy control systems, *Fuzzy Sets and Systems*, **14**, pp 199-210.

Tong, R.M. (1985), An annotated bibliography of fuzzy control, in *Industrial Application of Fuzzy Control* (M. Sugeno ed.), Amsterdam: North-Holland, pp 249-269.

Turksen, I.B. and Zhong, Z. (1988), An approximate analogical reasoning approach based on similarity measures, *IEEE Trans. Syst. Man Cybern.,* **18,** pp 1049-1056.

Vishnoi, R. and Gingrich, K.J. (1987), Fuzzy controller for gaseous anaesthesia delivery using vital signs, *Proc. IEEE 26th Conference on Decision and Control*, pp 346-347.

Voss, G. *et al.* (1987), Adaptive multivariable drug delivery control of arterial pressure and cardiac output in anesthetized dog, *IEEE Trans. BME,* **134,** pp 617-623.

Wang, G.J. and Miu, D.K. (1990), Unsupervising adaptation neural-network control, *Proc. Int. Joint Conf. Neural Networks,* pp III421-428.

Wang, L. (1992), Fuzzy systems are universal approximators, *Proc. of Int. Conf. on Fuzzy Systems.*

Wang, Li-Xin (1994) *Adaptive Fuzzy Systems,* Englwood Cliffs, NJ: Prentice Hall.

Werbos, P.J. (1990a), Neurocontrol and related techniques, in *Handbook of Neural Computing Applications* (A. Marsen *et al.* eds), London: Academic Press, pp 345-380.

Werbos, P.J. (1990b), Back-propagation through time: what it does and how to do it, *Proc. of the IEEE,* **78,** pp 1550-1560.

Werntges, H. W. (1990), Delta rule-based neural networks may be applicable to a control task that does not provide a "teacher", in *Parallel Processing in Neural Systems and Computers* (R. Eckmiller *et al.* eds), Amsterdam: North-Holland, pp 435-438.

Wong, Y.F. and Sideris, A. (1992), Learning convergence in the Cerebellar Model Articulation Controller, *IEEE Trans. Neural Networks*, **3,** pp 115-120.

Yager, R.R. (1984), Approximate reasoning as a basis for rule based expert systems, *IEEE Trans. Syst. Man Cybern.,* **14,** pp 636-673.

Yamaguchi, T., Takagi, T. and Mita, T. (1992), Self-organizing control using fuzzy neural networks, *Int. J. Control,* **56,** pp 415-440.

Yamashita, Y. *et al.* (1988), Fuzzy control of blood pressure by drug infusion, *J. of Chemical Engineering of Japan*, **21,** pp 541-543.

Yamazaki, T. (1982), *An Improved Algorithm for a Self-organizing Controller*, PhD thesis, QMC, University of London.

Yamazaki, T. and Sugeno, M. (1985), A microprocessor based fuzzy controller for industrial purposes, in *Industrial Applications of Fuzzy Control* (M. Sugeno ed.), Amsterdam: North-Holland, pp 231-239.

Ying, H. *et al.* (1988), Expert-system-based fuzzy control of arterial pressure by drug infusion, *Medical Progress through Technology,* **13,** 203-215.

Zadeh, L.A. (1973), Outline of a new approach to the analysis of complex systems and decision processes, *IEEE Trans. Syst. Man Cybern.,* **3,** pp 28-44.

Zadeh, L.A. (1978a), Fuzzy sets as a basis for a theory of possibility, *Fuzzy Sets and Systems,* **1,** pp 3-28.

Zadeh, L.A. (1978b), PRUF a meaning representation language for natural languages, *Int. J. Man-Machine Studies,* **10,** pp 395-460.

Zadeh, L.A. (1983), The role of fuzzy logic in the management of uncertainty in expert systems, *Fuzzy Sets and Systems,* **11,** pp 199-227.

Zadeh, L.A. (1989), Knowledge representation in fuzzy logic, *IEEE Trans. Knowledge and Data Engineering,* **1,** pp 253-283.

# Index